RENOIR

SELF-PORTRAIT. 1910. *Collection Durand-Ruel, Paris*

PIERRE AUGUSTE
RENOIR

INTRODUCTION BY

WALTER PACH

THAMES AND HUDSON

First published in Great Britain in 1984
by Thames and Hudson Ltd, London
Reprinted 1989

This is a concise edition of Walter Pach's *Renoir,*
originally published in 1964

Printed and bound in Japan

CONTENTS

PLATES

RENOIR

1. THE JUDGMENT OF PARIS. Collection Dr. and Mrs. Harry Bakwin

When I first visited Renoir in 1908, he was sixty-seven years old. I had come to him in the hope that he would grant me an interview, but the prospects of inducing him to talk were less than encouraging. He had not spoken for publication since 1879—some thirty years—when his brother, hoping to help him through a period of struggle and obscurity, wrote a piece about the artist and his work.

My visits with Renoir lasted over a period of four years; but instead of the reticence I expected to find after his years of silence, I found him eager to talk, freely and unreservedly, about art.

"There are days," he warned me, "when I chatter along like a magpie, and others when I can't say a word." Indeed, sometimes he would run on for so long that Gabrielle, who was both his housekeeper and his model, would rattle the dishes as a signal for me to be off. She feared that the conversations, after a full day of painting, would tire him. Frequently during the day, he was interrupted by terrible attacks of arthritic pain, some of them lasting as long as ten minutes. He would wait for the pain to subside, and then, his flow of ideas unbroken, he took up his conversation at the point where he had left it. Usually he spoke of his work, or of the painters whom he admired, but often he would return to memories of his youth. His story was a simple one, for his real career was in his work.

Pierre Auguste Renoir was born at Limoges in 1841, the son of working-class parents. In 1845 the family moved to Paris. Coming as he did from a city noted for its ceramics, the father apprenticed his son at the age of thirteen to a Parisian factory where porcelains and earthenware were made. Long before the end of the four years he spent at this establishment, the young Renoir, by virtue of a hand extraordinarily skilled, had become one of its most valued decorators.

2. THE PROMENADE. Private collection.
Photo M. Knoedler & Co.

It was during this period that he had his first introduction to great art, for he spent many hours at the Louvre, drawing from the antique. He later told Ambroise Vollard that Boucher's *Diana at the Bath* was his first enthusiasm among paintings; even after he came to know greater masters he still regarded Boucher as his "first love." During these years, Renoir chanced to bring his lunch one day to the square where Jean Goujon's sculptures adorned the Fountain of the Innocents. He was filled with admiration for these glorious evocations of the rivers of France, represented by strong, graceful, and carefree women. Certainly the terms with which he described this sculpture can be applied to his own mature art: "purity, naïveté, elegance, and, at the same time, solidity of substance."

At an early age Renoir had already decided that painting was his true career; and with the money he managed to save from his commercial employments he began his studies in the art schools. Years of poverty followed; at the age of thirty, he was still grateful for a full meal at his father's table. The scraps that were left he would bring to his friend Claude Monet, who was often in still greater need.

Though he served in the army during the war of 1870, Renoir's share in the struggle was a minor one. The real battles of his life were those of his profession. Few schools have met such opposition as the Impressionist, and Renoir was one of its leaders from the beginning. He was active in organizing the group's exhibitions and sales for twelve years; it was only gradually that, during this time, the hostility of the critics and public began to be overcome. Renoir never wavered in his admiration for early comrades like Monet, Cézanne, and Degas, but in 1886, after dissensions in the group had changed its character, he refused to continue his participation in their shows. He had given ample proof of his solidarity with the

3. THE MISSES LEROLLE. Collection John S. Newberry, Jr. *Photo Jacques Seligmann & Co., Inc.*

men who made the great artistic advances of their time; but, like Monet, he understood also that the essentials of art stood outside the doctrines of any single school.

Able to work quickly and to catch the likeness of a sitter, Renoir gradually attained a measure of financial security. In this he was greatly aided by the heroic support of Paul Durand-Ruel, the dealer who had espoused the cause of the Impressionists.

When, in 1881, Renoir for the first time had solid sums of money in his pocket, he fulfilled an ardent desire to travel and to make further study of the masters. He visited Italy and Algeria;

and on his return from this journey, he married one of his models, Aline Charigot, whom he had painted in *Luncheon of the Boating Party* (page 89). Madame Renoir was devoted to her husband and the three sons she bore him, and their household was one of deep harmony.

In the eighties began the ill-health which was to dog the artist throughout his life and which stands in such striking contrast to the joyful quality of his painting. He suffered frequent periods of intense pain. Yet, never was his art so radiant as in his later years when he was deprived of the use of his legs, and, unable even to bring thumb and forefinger together, he had

4. GABRIELLE AND JEAN. Private collection.
Photo M. Knoedler & Co.

to have his brushes fastened in his hand with a kind of clasp.

In 1915 Madame Renoir died, after the overwork and strain of nursing back to health the two sons who had been severely wounded in the First World War. The monument for her grave is adapted from Renoir's sculpture of his wife and their infant son, Pierre. Her death was a tragic blow to Renoir; but nothing, not even death, could dim the inexhaustible joy of his painting. The very craft and technique of his art were a constant delight to him. And yet,

with all his devotion to the "trade" of painting, Renoir never thought of it solely as a matter of skill; on more than one occasion he spoke with contempt of pictures characterized only by manual dexterity. For him, there was lacking in such merely facile work a recognition of the supremacy of the mind. And that conviction not only remained, but was stronger than ever in his last years, when the use of his hands was so nearly gone. In 1919, on the day before he died, he said, "I am still making progress." And at the moment of his final weakening, as he directed the arrangement of a subject he still wanted to paint, the last word he said was "flowers," surely as near to his life work as any he could have used. The impression that Renoir had made on his contemporaries through both his character and his art was summed up by Elie Faure, who wrote of his death, "It is as if the sun had gone out of the sky."

The beautiful things of nature were the great stimulants for his art, and when painting the figure, he never worked without a model. But behind his apparently simple love of nature stood an exalted conception of art and tradition; and this distinguished him from a thoroughgoing Impressionist like Monet, whose paintings concentrated—but how wonderfully—on momentary visual effects. Throughout his career, Renoir was deeply devoted to the great works of the past and from them drew important inspiration for his own original thinking and creation. Thus the boyhood discovery of the lovely sixteenth-century sculpture of Jean Goujon, already mentioned, is surely related to certain figure-pieces which he painted more than forty years later. Such devotion, not unique in his career, furnishes the key to his emphasis on tradition and makes it easier to understand his rejection of the Impressionists' absolute and exclusive submission to nature. As he expressed it, "With nature one is bound to become isolated; I want to stay in the ranks" (a French dictum if there ever was one). To the question of the German critic, Meier-Graefe, "Where then should one learn to paint?" Renoir replied, "In museums, *parbleu!*" Even in 1919, the year of his death, Renoir—unable to walk but eager to see once again the masterpieces he loved—traversed in his wheel chair the long galleries of the Louvre to which the paintings had been returned from their wartime hiding place.

When, in his youth, he worked in the art schools, the "grand style," then being taught, always was more to him than the mere byword which it was to his friends Monet, Sisley, and Bazille; they ridiculed it in their devotion to the down-to-earth realism of Courbet. Renoir himself drew important ideas from Courbet (whom even so visionary an artist as Odilon Redon has called a classic master). But while an almost naïve acceptance of reality was all that the majority saw in the great mountaineer of the Franche-Comté, Renoir, like Redon, saw in him a stupendous solidity of form and magnificently balanced design. Such qualities gave our painter his confidence in the realist's frank reliance on the object as seen, the feeling that Courbet transmitted to all the strong young men of his time.

But while Renoir shared this directness of vision with his fellow Impressionists, they did not share in another enthusiasm of his formative years, one which was to have a great influence on him— the work of Ingres. In the historic rivalry between Ingres, the Classicist, and Delacroix, the Romantic, a rivalry which dominated the aesthetic polemics of the mid-nineteenth century, the Impressionists were naturally partisans of Delacroix, for them the patron saint of color, and against Ingres, who gave priority to line and clear form. Renoir, too, adored Delacroix, some of whose paintings he adapted, but he could also see the

beauty of Ingres' pure severity and classic discipline. While copying a Delacroix in the Louvre, he could not withhold his admiration for Ingres' *Portrait of Madame Rivière*, which hung nearby. He also loved, as he said, the little belly of the figure in *La Source*, by the same master. Renoir was by nature a composer, a painter of the human figure. As such, he faced problems of drawing and modeling which never confronted his landscape-painting comrades. He could therefore more readily grasp the virtues of the severe Ingres. It took courage for him to admire that stubborn enemy of modern color; and in this heretical taste he gave another sign of that genius for form which, as time went on, distinguished him more and more from every Impressionist except Cézanne.

Continued on page 24

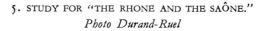

5. STUDY FOR "THE RHONE AND THE SAÔNE."
Photo Durand-Ruel

Painted 1874

BALLET DANCER

56⅛ × 37⅛"

National Gallery of Art, Washington, D.C. (Widener Collection)

Unlike Degas, who sees the ballet dancer as an object of the theater, a trained performer, Renoir here presents the little dancer more sympathetically as a charming girl. The ballet is formalistic—traditional and aristocratic; and this combination would naturally appeal to Renoir and Degas. They were the two among the Impressionist group who, though concerned with what was immediately before their eyes, still retained a great respect for formal composition and tradition.

Renoir gives us here the graceful poising of a figure destined for beautiful motion. Indeed, the movement, weightlessness, and fleeting balance of the ballet may be called the unseen subject of this painting. Cunningly, every detail is made to contribute. The dancer is suspended in a cool, relatively high-key atmosphere which suggests floor and background, yet does not engage the figure in architecture. There are no hard boundaries; the drawing is delicate and the touch light—and yet there is precision and fluidity.

Through the dainty attractions of the extremities of the body, our attention is made to shift between widely separated areas of lovely decoration: the details of the neck, face, and hair; and the legs and feet. The brightest touches of color in this picture—an arrangement of pearly tones on grayish tones—are very diluted spots of red, yellow, and blue, and Renoir's black. Within the restricted color in each area, there is an amazing variety of tints, and the brushing at times seems to dissolve into air.

There is a small series of triangular notes, like the one between her feet, and above it, between the ankles; two more triangles of shadow at the shoes; the triangular arrangement of arms and head; the triangle of the projecting skirt. One fancies that something of the lightness of the piece may derive from the suggestion of a wing in the drawing of the skirt. The lightness, the delicacy, the fluttering life of a creature not earth-bound, are given to perfection here.

Painted about 1875

PORTRAIT OF VICTOR CHOCQUET

18⅛ × 14¼"

Oskar Reinhart Collection, Am Römerholz, Winterthur, Switzerland

A kind of fame belongs to this otherwise obscure customs office clerk: he was friend and sitter to some of the greatest artists of his time. Passionately devoted to progressive art, Chocquet had seen Renoir's work at the historic Impressionist sale at the Hotel Drouot in 1875. He immediately wrote, begging him to do a portrait of Mme. Chocquet with one of Delacroix's pictures in the background. "I want the two of you together, you and Delacroix." Subsequently Renoir painted Chocquet twice. That the painter was touched by the man's eagerness, sensitivity and gentleness is obvious in this affectionate portrait, painted at a time when Renoir was sorely in need of a patron, and beset by financial problems.

A fine, high-keyed tonality and a lightness of brushing mark Renoir's authentic manner of this period, differing from the tighter rendering of his society portrait commissions. In this relaxed portrait, there is a wonderful combination of sympathetic understanding of the sitter's personality and a graceful skill that harmonizes perfectly with it.

Cézanne painted Chocquet, too; in his version one is struck immediately by Cézanne's rugged, almost fierce modeling, his preoccupation with structure, the bones beneath the flesh. Renoir, captivated by his sitter's personality, reveals how much more important to him was the projection of the spirit of the man (in the last analysis, when portraitists are at their best, they are painting themselves as well as their sitters).

One sees here, in the casual, almost untidy painting of the hair and shirt, still more of that interest in the informal that produced Renoir's most intimate portraits. In the delicate, free painting of the fingers, whose outlines are inexact, there is much of the instinctive and tentative; Renoir's true interest was in things as experienced and felt.

Out of meager financial resources, Chocquet bought canvases from Renoir, Cézanne, and the other Impressionists before their work was widely appreciated. He strenuously fought for the recognition of the young painters, and managed to build a magnificent collection of their works. The collection was dispersed after his death.

Painted 1910

PAUL DURAND-RUEL

25¾ × 21½"

Collection Durand-Ruel, Paris

A hundred years ago, in Paris, there were no dealers' galleries such as exist now, with their regularly changing exhibitions, their publications, their "stables" of artists; and there were only a few shops where paintings could be seen. Artists had to rely for professional income on chance sales, the hope of public commissions, teaching assignments, and luck. For the progressive painters the road was hard.

Shortly after the Paris World's Fair of 1885, Paul Durand-Ruel, a dealer in paintings, opened a shop in the Rue de la Paix. He had been fighting for recognition of the Barbizon painters, whose landscapes now seem so Arcadian in their peace and intimacy; and soon he was to have more virulent battles: he became a champion of the Impressionists. He began buying their works around 1870, and from then on the fortunes of the Impressionists and of Durand-Ruel were interlocked.

In 1866, still buffeted by critics, he took the advice of an American friend and opened a show of choice pictures in America. As he recalled it in his memoirs: "The exhibition was a great success, invited curiosity, and the reverse of what happened in Paris. It raised no hubbub, and not a single protest. The praise was unanimously favorable, and numbers of articles praising the show appeared in the New York newspapers, as well as in all the large cities." As a result of the favorable reception (which was actually not quite as "unanimous" as here recalled), he opened the Durand-Ruel Gallery in New York.

It was largely through this man's efforts that Renoir's—and other Impressionists'—financial problems were eased, and he thus made for himself a place in the history of modern painting.

Painted about 1882

THE STAIRWAY, ALGIERS

28½ × 23½"

Collection Mr. and Mrs. Grover Magnin, San Francisco

This painting of torrid sunlight and dense foliage shows us much about Renoir's way of approaching a subject. Albert André writes: "He attacks his canvas, when the subject is simple, by putting in with the brush, usually in brownish-red, some general outlines in order to see the proportions of the elements which will compose his picture... then, immediately, with pure colors diluted by turpentine, as if he were painting with watercolor, he strokes the canvas rapidly and soon you see something vague appear, with iridescent colors, the tones flowing into one another; something that charms you even before you grasp the sense of the image."

The figures in the extreme right foreground are lightly indicated by a few strokes of the brownish-red which establishes the general scheme of the picture. With this shorthand note as a clue, one becomes aware that a major element of the compositional organization occurs in the *degree of finish* of the various parts of the canvas. In the figures and steps of the foreground there is a sketchy, freehand quality much like the technique of watercolor. The loosest technique is in the most open areas; farther up, the brushwork of the wall and the trees becomes closer, the pigment thicker, and the foliage richer in detail. The separated color areas of the foreground are developed with a great range of hues at the top of the canvas.

Renoir delights in the contrast of the sunlight, which bleaches color where it hits the scene directly, and the glow of the shadow areas. Shadow here is not darkness—the absence of light—but another kind of light and color; it is hard to imagine a more wonderful, varied play of related colors than the iridescent light and warm lavenders of the wall and façade at the top. Something of each of the different elements—the small brush notes of the foliage, the architecture, and the softly painted sky—is fused in the treatment of the buildings. The human figures are merely sketched in; yet Renoir's sure instinct for essentials gives them convincing reality.

Renoir's love of form showed itself in the supremely important visit he made to Italy in 1881, which has already been mentioned. He appreciated the color of the great Venetians, but his development at this time was more decisively influenced by his contact with Raphael and the masters of fresco painting. Prepared for Raphael by his study of Ingres, Renoir asserted that it was the latter artist whom he preferred in easel painting, but that Raphael, in his frescoes, showed himself to be the greater man. The composition of the Italian was of course splendid, but not less impressive was the beauty attained by working directly on plaster walls. The sense of clarity and

6. DANCER WITH TAMBOURINE. *Photo Durand-Ruel*

true decoration achieved in this kind of painting was of increasing importance to Renoir, who tried constantly to infuse these same qualities into his canvases.

At Naples, where an earlier classical art, that of Pompeii, excited his admiration, Renoir began a new series of studies. The nude figure is never so natural and resplendent as under the southern sun and profiled against blue Mediterranean waters. Here the artist cleared from his paintings of the nude the last of that "brown sauce" which the academicians in Paris used in their modeling of form. The Impressionists had banned from their palettes a dark tarry substance called bitumen, formerly used to obtain a golden glow; in its place they worked with color, color everywhere, even in the deepest darks. And they achieved a luminosity in their landscape painting such as the world had not seen before. Applying the new control of light-through-color to the richer problem of the human figure, Renoir, first at Naples and then in Paris, painted a great series of nudes in the decade beginning in 1881. In the limpid Florentine frescoes and in Raphael's orchestration of forms he had just seen the noblest Renaissance solutions of the problem of composing with human figures; and Naples offered further revelations through the still-sculptural conception of the later Greeks.

We feel the naturalness with which Renoir accepted these influences in a sentence of his about the art of Pompeii, of which the Louvre then had no good examples to prepare his mind: "It was Corot himself, the whole of his art, that I found again at the museum in Naples, with that simplicity of workmanship that the Pompeians and the Egyptians had. These priestesses in their silver-gray tunics—you would absolutely think you were seeing the nymphs of Corot."

7. STUDIES OF HEADS. Collection Robert Lebel

From Naples, Renoir went on to Sicily. In Palermo he obtained a brief sitting from Richard Wagner. Like others of the elite in Paris, he had been excited by the German composer's music, but his preference was for the French masters, who, though less grandiose, seemed more natural to him.

In this period Renoir also visited Algeria. Its sun and its mystery reawakened his love for Delacroix, whose sojourn in North Africa fifty years before had resulted in an enormous development of his sense of color. But whereas Delacroix's work was deeply influenced by his Algerian visit, there was no comparable change in Renoir's. How-

ever, like his great predecessor, Renoir, upon returning to France, began to prepare a masterpiece which would unite his new sense of light and color with the deepened conception of form which he had developed during his travels.

Italy continued to fascinate him after he had returned to Paris. The nudes he had done at Naples were given new renderings, with the contours drawn in severer, more sculptural outline. It was the beginning of Renoir's *période aigre*—the sour period—so called by those who saw in this evolution only the loss of *morbidezza*, that softened mystery which is characteristic of Impressionist painting. A typical critical attack was by George

25

Moore, the Irish novelist, who had defended Manet, Degas, and their comrades; and yet he wrote that Renoir "in two years had utterly destroyed the charming and delightful art which had taken him twenty years to build up." Moore, a sensitive appreciator of Renoir's earlier period, made the common blunder of condemning a new development because it was different from its past, a past which the artist had outgrown. Had Renoir died in middle age—when he had already painted many remarkable pictures—the world would not today feel that an abyss separates him from various of his contemporaries, already half-forgotten. The masters themselves know best

8. HEAD OF A CHILD. Private collection

what their course should be, and that Renoir continued to progress despite the physical infirmities of old age, is borne out by the superiority of his later work—even when compared with such masterpieces of the seventies as *Moulin de la Galette* (page 67), or *Madame Charpentier and Her Children* (page 77).

In these two paintings Renoir had not yet left his period of schooling. He himself characterized it as "the time when I was an Impressionist." The turning point came shortly after his return from Italy and Africa—with the onset of the *période aigre*. Years later he was to tell Vollard of this crisis in which he grew increasingly uncertain of Impressionism and began to despair of his own art.

"About 1883 I had wrung Impressionism dry, and I finally came to the conclusion that I knew neither how to paint nor how to draw. In a word, Impressionism was a blind alley, as far as I was concerned....

"I finally realized that it was too complicated an affair, a kind of painting that made you constantly compromise with yourself. Outdoors there is a greater variety of light than in the studio, where, to all intents and purposes, it is constant; but, for just that reason, light plays too great a part outdoors; you have no time to work out the composition; you can't see what you are doing. I remember a white wall which reflected on my canvas one day while I was painting; I keyed down the color to no purpose—everything I put on was too light; but, when I took it back to the studio, the picture looked black.... If the painter works directly from nature, he ultimately looks for nothing but momentary effects; he does not try to compose, and soon he gets monotonous."

Such doubts and difficulties led him to search for a new method and (with his experience of Italy fresh in mind) he turned to the lessons of

9. STUDY FOR "THE BATHERS." Private collection. *Courtesy M. Knoedler & Co.*

Ingres and the classic Italians. The influence of Ingres is evident in certain felicities of design, but Renoir's course, though traditional, was not one of slavish imitation. The picture on which his art focused in the first few years of the eighties, the canvas which the painter himself at that time declared to be his masterpiece, is *The Bathers* (page 99), in the Tyson collection, Philadelphia Museum of Art. Its former owner is authority for the statement that among the drawings used for the work was one that Renoir had made of a late seventeenth-century relief by Girardon at Versailles. One figure in this relief, that of a girl splashing water at a companion, is repeated in *The Bathers*. Again we have an indication of Renoir's love of that tradition in French art which produced Watteau, the master with whom he has perhaps the most marked affinity. In the Watteau of the lovely *Fêtes Champêtres* and the sober *Signboard of Gersaint*, as in Renoir, the qualities of elegance, naturalness, and charm which immediately appeal to us are discovered to have design of almost architectural logic for their secure foundation.

Renoir was undoubtedly engaged in developing this splendid sense of order as he experimented for three years with a series of studies which he varied so often in preparation for *The Bathers*. His love for the drawing of Ingres and the lyrical genius of Watteau relates this picture to the past; but far more important is its prophetic quality, its anticipation of the work which was to follow. Once he had fought out the detailed rightness of

10. EDMOND RENOIR AT MENTONE. Private collection

design that is here, the response of every line and mass and color to every other element in the picture, Renoir continued to apply this high standard to all his paintings. The slightest, most spontaneous sketch from nature would profit by his new advances quite as truly as did the larger, more considered works of the succeeding years.

During the years he spent on *The Bathers* he turned aside to paint many other pictures, and in some of these we find qualities which distinguish the masterpiece, brought to even greater perfection. Thus, in the several versions of *Madame Renoir and Her Son Pierre*, there is a very direct technical effort to capture the beauty of fresco, and a stronger emphasis on form. Only Cézanne, among the Impressionists, showed a comparable solidity of construction.

These paintings of the *période aigre*, to return to that term of misunderstanding, have many suggestions of the purity of Clouet and his use of clear contours and unshadowed color. Or, one is reminded of that first typically French school of painting which flowered around the palace of Fontainebleau in the sixteenth century; in that group, as with Renoir himself, Italian genius had influenced French art, adding strength and modeling to warm color and acute observation. In the following decades, as Renoir's mastery increased, the lessons of a great technician were added to those of the composers and constructors he had studied before; Velázquez became a chief admiration, and Renoir would point out the Spaniard's way of opposing heavy dabs of pigment to the thin wash of color which began the painting.

In his respect for Velázquez, we observe again Renoir's willingness to learn from the masters. He once declared to me that tradition has never been an obstacle to originality, and he used a supreme artist as an example: "Raphael was a pupil of Perugino, but that did not prevent him from becoming the divine Raphael." This faith in tradition is of especial value today when few students copy the old masters—one of their chief means of learning from the past—and when there is a widespread notion that modern art is essentially different from the art of the past and may even suffer by contact with it.

Renoir showed a lifelong devotion to Rubens: in his old age, returning from a trip to Munich, he overflowed with the wonder that had filled him before the sketches and finished work of Rubens at the Alte Pinakothek. Delacroix's observation that Titian is the master whom old painters best appreciate also held true for Renoir. His later pictures, in which a general ruddy flesh tone is varied by insinuations of opaque, whitish colors, offer a striking homage to the genius of the great Venetian who produced his final and finest painting in the same fashion. Renoir made no secret of his debt and spoke of it with a jest: "That old Titian! He even looks like me, and he is forever stealing my tricks."

It was Elie Faure's insight that penetrated to a quality in Renoir's last period which usually escapes notice. Faure pointed out the relation to Michelangelo, whom Renoir had, to an extent, resisted during his stay in Italy. The painter had criticized a certain too-great similarity among the figures of Michelangelo, a too-complete acquaintance with anatomy "despite all of his genius." But in the last year of his life, Renoir said: "And Michelangelo himself, the anatomist par excellence! The other day I feared that the nipples of my big statue, the *Venus*, were too far apart; and then I just stumbled on a photograph of the *Dawn* on the tomb of Giuliano de Medici, and I could see for myself that that very Michelangelo didn't trouble his mind a bit about having put an

II. YOUNG GIRL WITH A ROSE. *Photo Durand-Ruel*

even bigger space between the two breasts." Who knows what far-off memories of his visit to the Medici Chapel in Florence, a quarter-century or more earlier, may have influenced him in his own masterpiece of sculpture? In any case, after Elie Faure made that observation, it was clear that the heavy limbs and bodies of Renoir's late pictures, their heroic cadences, and their reaching beyond merely gracious and charming gestures marked a development analogous to that of the later Michelangelo, in rising above his *Pietà* at St. Peter's or the first *David*.

Continued on page 38

Painted 1906

LADY WITH A FAN

25⁵/₁₆ × 21¼"

Private collection, Switzerland

How sensitive an instrument Renoir was impresses us as we turn to this full-blooded picture of a woman. Every aspect of the plate opposite tells of robust health and comfortable circumstance. Incidentals which would distract from the self-assurance and placidity of the subject are omitted. Instead, we have full, regular curves in untroubled harmony, full volumes sonorously echoing one another, simple color areas of cello-like richness, a broadened pose which gives the painting a stately splendor. In keeping with these qualities, Renoir's decorative enrichment is held to a minimum: the fluting of the fan is noticed and the ridges are repeated in the drapery in the upper right corner and in the column-like arrangement at the left; the flowers in the hair and on the dress are sketched in; the transparent color of the sleeves is relieved by small dots. Renoir responds completely to the requirements of his artistic experience of his subjects, and style and technique are in perfect accord with the vision.

Painted about 1907

GABRIELLE IN AN OPEN BLOUSE

25¾ × 21"

Collection Durand-Ruel, Paris

In this painting Gabrielle is not represented as a strongly individualized personality: what Renoir has painted can stand for everything sweet and artless in maidenhood which is not yet quite conscious of its womanliness. The picture seems inevitable and right; there is nothing labored, nothing forced. The lively brushing of pearly grays and satiny whites gives the blouse the effect of a translucent cocoon, from which emerges the beautiful pink torso. The climax of color is in the face, broadly modeled and set off by the charming simplicity of the hairdress, the dark flow of which is contrasted with the dainty blossom.

In the background there is nothing to distract from this delightful vision; instead, there is a field of turquoise blues, wonderfully varied in hue and texture. In a picture like this, we see a great artist in a relaxed mood, his inspiration pure, spontaneous, and free from any suggestion of artifice.

Painted about 1914

THE JUDGMENT OF PARIS

38 × 46"

Hiroshima Museum of Art

The act portrayed here led to war. The shepherd-prince, Paris, a famed connoisseur of beauty, was called upon to give a golden apple marked "For the Fairest" to the most beautiful of the three competing goddesses, Juno, Minerva, and Venus. As though the sudden appearance before him of three nude goddesses were not enough, each one offered him bribes: power and riches, renown in war, or the fairest of all mortal women as wife. The last was understandably not least, and Paris did not hesitate to accept, at once awarding the golden apple to Venus. The "fairest of all mortal women" turned out to be Helen of Troy, and her subsequent elopement with Paris is what, in mythology, set the Greek and Trojan armies at each other's throats.

Renoir shows us the moment of the bestowal of the apple, as Mercury, at the left, signals the end of the contest.

The picture has a curious, but disarming naïveté. Paris, wearing a shepherd's robe and Phrygian bonnet, is Gabrielle, as is the goddess at the right. The landscape is in the Impressionist manner, though it represents a classical scene. The nudes are typical of Renoir's late period in their small-breasted, long-waisted, big-hipped design, statuesque and voluminously modeled, but painted in soft, flowing colors; the faces have nothing of the traditional idealized classic type.

Renoir is untroubled by the fundamental contradictions of subject and style, and indeed, one of the striking things about the picture is the degree of success with which he resolves them into an original unity. He has not fallen into the academic trap of pretending he was more-Greek-than-the-Greeks; it is not the outward appearance of the classical tradition—worn out by centuries of misuse—which has motivated him, but its essential spirit. Thus, without self-consciousness, drawing from his own ever-youthful vitality, Renoir brings new life to the great humanistic tradition.

34

Painted 1918

GIRL WITH A MANDOLIN

25⁵/₁₆ × 21¼"

Collection Durand-Ruel, Paris

A comparison of this plate and the one on page 127 gives insight into the workings of the painter's mind at the end of his career. Famous and wealthy, recognized throughout the world as an artist of stature equal to that of the great painters of the past, Renoir was still painting with the joy and excitement of a young man who has just found his career. When a theme interested him—and everything carefree and pleasant did—he was content to paint it over and over again, for out of his own inner resources he could produce endless original variations.

The two plates show the same activity, similar props, and resemble each other in the breadth of the drawing and in other ways. Yet the present plate has, in general effect, Renoir's blonde, silver-gold tonality, whereas the other has more striking contrasts of color. The very shapes of the canvases have determined to some extent the quality of the pictures: this one seems freer in action and more serene, the lines and surfaces more flowing; while in the other plate we are aware of a greater density of forms, a heavier texture, and a certain compression.

In both pictures—small ones, by the way—evidently painted in the same corner of the studio, we find a blue area in the upper left-hand section, treated quite differently in each case. The wallpaper has a design of roses; in the present plate, Renoir has wittily set up an evenly spaced continuity between the patterned roses and the climactic real one in the girl's hair. The colors of the wall are also the colors of the diaphanous blouse; and the burnt almond of her skirt, used as an edging, helps to separate these color areas, while at the same time repeating the curved, linear motif. Against this overall similarity of hue and tonality Renoir singles out the deep pink flower, the yellow of the mandolin top, and the blue in the corner for solo appearances. The dark notes in the very extreme right-hand corner, the mandolin, and the hair are sharply in contrast with the general filminess. As the eye goes from the dark spot in the corner to the darks of the mandolin and then to the hair, its motion is that of a long arc—and here again, we see how subtly Renoir relates elements of his picture.

This massive quality in Renoir's painting is shown most clearly in that miracle of his old age, the sculptures. A certain mystery still surrounds them: Paul Haesaerts' volume on these works does not fully explain how they were produced, and some critics have even gone so far as to doubt that they can be called Renoir's. The artist's oldest son, Pierre Renoir, who was a most distinguished figure in the French theater, watched his father work on the sculptures, and he gave me an account of their genesis which is unquestionable in its accuracy. This extraordinary story, which I have related before in another book, removes any doubts about Renoir's authorship or his methods. It is of the later sculptures that the son spoke, for at least two of the earlier figures Renoir produced unaided; they date from a time

12· NUDE WOMAN DRYING HERSELF. *Photo Durand-Ruel*

(1907—8) when he could still use his hands. Later, when his arthritis made the physical effort of sculpture impossible for him, he was obliged to employ assistants.

The first was a young Spaniard named Guino, a pupil of Maillol, and recommended by the latter to the great painter he so much admired. Renoir's sketch on paper or canvas would furnish the general setup of the mass of clay. With a light wand the old man would indicate how the work was to continue: at one place making a mark on the clay sketch where the ready modeling tool of the young sculptor was to cut away an exaggerated volume, and at another place showing by a pass through the air how the contour was to swell out through the addition of more clay, which the nimble fingers of the assistant molded until the desired form was achieved. Then the figure would be turned on its pedestal and would need trimming down to compensate for an excess invisible from the other angle. The wand would again travel down the surface of the clay, the tool in the hands of the young man following and scraping to the desired depth.

"They communicated by grunts when the thing got so close to the desired result that it grew exciting," related Pierre Renoir in describing the scene. "My father would say, 'Eh, eh, eh-ahh! *Ça y est*,' and the sculptor would be making his little jabs with the tool, so much like my father's brush strokes that you'd think the whole thing came from his hand, as indeed it almost did; and they would be working together so closely that the young chap would say, 'Eh, eh? Eh?'— only with a note of interrogation that made you feel he was responding to those other grunts and did not propose to go too far by the thickness of a cigarette paper."

Three later sculptures were executed by another assistant, Morel, who lacked the self-effacing quality of his predecessor; these works do not therefore so completely render the intention of the master. Yet seldom has a painter been so successful in transferring to sculpture the spirit of his canvases. It is fitting that the pieces are signed by the man who gave them to the world.

Some of Renoir's statements about art throw important light on both the man and his work. In one of his conversations with me he gave a supremely simple account of his method and aim as an artist, and this is so much more revealing than any critic's observation that I cannot refrain from quoting it again.

"I arrange my subject as I want it, then I go ahead and paint it, like a child. I want a red to be sonorous, to sound like a bell; if it doesn't turn out that way, I add more reds and other colors until I get it. I am no cleverer than that. I have no rules and no methods; anyone can look at my materials or watch how I paint—he will see that I have no secrets. I look at a nude; there are myriads of tiny tints. I must find the ones that will make the flesh on my canvas live and quiver. Nowadays they want to explain everything. But if they could explain a picture, it wouldn't be art. Shall I tell you what I think are the two qualities of art? It must be indescribable and it must be inimitable.... The work of art must seize upon you, wrap you up in itself, carry you away. It is the means by which the artist conveys his passions; it is the current which he puts forth which sweeps you along in his passion."

To read Renoir's words is to have a new confirmation of his belief that art is timeless, that what was right once is right always. He expressed that principle in another conversation with me:

"There is nothing outside of the classics. To please a student, even the most princely, a musician could not add another note to the seven of the scale. He must always come back to the first

13. THE WASHERWOMAN. Buchholz Gallery
(Curt Valentin)

one again. Well, in art it is the same thing." With this went a warning to those who do not see that seven notes never reach the same combination twice. "You cannot make any more Titians, and you cannot copy Notre Dame. There is the Pantheon at Rome; they thought they could make a copy of it in that votive church at Naples, opposite the Royal Palace; but the Pantheon is a great thing and that church is a dead thing. Again, when they try to build like the Parthenon they find that those lines which seem so straight and regular and simple are very subtle and hard to follow. The more they measure, the more they realize how much the Greeks departed from regular and banal lines in order to produce their effect. So in our Gothic architecture: each column is a work of art, because the old French monk who set it up and carved its capital did

14. DANCE IN THE COUNTRY. Private collection

what he liked—not doing everything alike, which results when things are made by machinery or by rules, but each thing different, like the trees in the forest."

As Renoir once wrote me, he did not want "to talk like a professor." His own conversation was largely of everyday matters, often treated with the ironic humor that is so genuinely French; and it is thus we see him in the pages of Ambroise Vollard's brilliant reporting. Vollard's book, however, ends on a slightly misanthropic note, which is anything but typical of the painter, who, despite misunderstanding in his later life as well as in his youth, despite intense physical suffering that would have been unendurable for a less courageous man, retained to the end his blithe love of nature, human beings, and art.

The most authoritative words of Renoir, because they are signed by his hand, are those of a letter to Henry Mottez, whose father, the painter Victor Mottez, had translated Cennino Cennini's *Treatise on Painting*. The son was publishing his father's book with Renoir's letter as a preface. This letter, which helps us to understand both Renoir and the Italian writer, first became accessible in the review *L'Occident* for June, 1910. Renoir was commenting on the tradition of the artisan:

"One must insist on the fact that it is the ensemble of the works left by numerous artists, forgotten or unknown, which makes up the greatness of a country, and not the originality of a man of genius. The latter, isolated among his contemporaries, cannot be thought of, most often, as remaining within national boundaries or within a period: he transcends them. But those lesser artists incarnate both the period and the country, even to the quality of its soil. With this much said, and with no desire to underestimate the glory bequeathed to their age and their land by men like Raphael, Titian, Ingres, and Corot, one may at least say that it would be impossible to think of writing a treatise on painting for those exceptional beings. The men whom the Italian master addressed did not, in every case, possess genius, but they always remained marvelous workmen.

"Now, to make good artisans of his readers was the sole purpose Cennini had in mind: your father well understood the practical results of such a course.

"Victor Mottez, one of the favorite pupils of Ingres, had the same admiration as his master for those great schoolworks, in reality collective masterpieces, which characterize the Italian Renaissance.... He was well aware of the fact that the great decorative ensembles of the Italian masters are not the work of a single man, but are collective productions, those of a workshop animated by the mind of the master. It was such a collaboration which he hoped to see reborn, in order to produce new masterpieces."

The generous ideal of the Ingres pupil was to reappear in Van Gogh's vision of communal work among artists (for a short time before the outbreak of World War I, something like this was actually put into effect by Duchamp-Villon and his group). Cennini's own book, Renoir said, can explain why the project, in the time of Mottez, had to fail. For the painters of the Renaissance, the letter continues, "the glory of having realized a beautiful work took the place of pay; they labored to gain heaven and not to make money."

Without a thought of describing his own career and its underlying faith, Renoir, through his description of the great artisan class into which he was born, gives us a valuable insight into the nature of his own creations. Not only was he a marvelous workman, he must today be recognized by the world as a man for whom the accomplishment of artistic purpose "took the place of pay."

15. GIRL WITH A HAT. Collection M. Knoedler & Co.

Why is it that though he painted pretty girls, clothed or nude, graceful landscapes, fruits and flowers, Renoir never for a moment could be ranked among those artists who seek to please merely through the attractiveness of their subjects? He belongs in the company of his youth, the acrid Degas, the generous Pissarro, the austerely passionate Cézanne. Those who study more recent artists delight to keep Renoir on their walls with the men whose researches into the principles of art make them indifferent to nature as seen. This indifference, at first glance, would seem to be the antithesis of Renoir's song in praise of nature's beauties. But there is no more of a contradiction here than there is between Renoir, who never painted a religious subject, and those painters of the Renaissance whom he loved for works almost exclusively devoted to

41

religious themes. A Renoir *Venus*, in its pantheistic adoration, may well be as religious as many a Florentine or Umbrian Madonna. The Madonna is a very pretty lady with a baby, and marvelously painted, but perhaps her picture is less truly religious and less dedicated to the "glory of having realized a beautiful work," as Renoir expressed it, than many a secular picture, which, however, may well be religious in a pantheistic sense.

Renoir realized, as we have seen, that art may have the most various manifestations; he knew that to recognize new forms frequently requires a long time, during which the observer adjusts his vision to a conception which may at first have seemed strange to him. Since he had undergone years of hostile misunderstanding and had seen Cézanne endure it for even longer, he did his best not to allow any words of his own to cause injustice to a newcomer. Thus, when I asked him in 1908 for his opinion of Matisse, he practically declined to answer, pleading his old age and his unfamiliarity with the work of the younger generation. Hearing of this, one of his friends said, "That is just like Renoir: he will never say an unkind word about anyone—not even about Matisse, whom he can't bear." But, in 1918, his refusal to commit himself to a premature judgment of the radical innovator was rewarded by the satisfaction of realizing that art was safe in the hands of a strong new generation. In Matisse, he exclaimed, was "painting, good painting!"

Renoir's career was one of extraordinary continuity. Through his entire life he kept with him those bits of decoration on porcelain which he had made as a young workman; they served as reminders that he himself, the famous artist, had begun as a member of the craftsman class, and was heir to its traditions. He meant high praise when he said that Cézanne, in his use of color, resembled the old masters of ceramics.

As we leaf through the illustrations to this volume we shall notice how the simple realism of Renoir's Courbetesque beginnings developed into a further study of nature when, with Monet and the other Impressionists, he made an intensive analysis of light and its rendering by color. During his next stage, under the impact of masterpieces seen at Florence, Rome, and Pompeii, and his new realization of the possibilities revealed to him by Ingres and Delacroix, a steady and intelligible transition from his Impressionist period took place. A purely Impressionist painting like the *Girl Reading* (page 69) is surely a sister work to the *Girl with a Straw Hat* (page 97) of the *période aigre*. The later years with their increasing opulence and freedom are in turn a gradual unfolding of what had preceded. Even the sculpture, that astounding innovation of the last decade of his life, can be seen as a thing long prepared for in his mind. An instance is the bronze which repeats, after a quarter of a century, the painting *Madame Renoir and Her Son Pierre*, of 1885. Renoir's sculpture was not only anticipated by his powerful rendering of three-dimensional form on canvas, but was also foreshadowed long before by his boyhood discovery of Jean Goujon.

His vision was comprehensive, and he judged Reims and Notre Dame, the cathedrals of his own country, in the same way that he judged the Parthenon. To Renoir's mind, the arts were one; thus a good picture must contain the order and logic of architecture, the meeting of physical reality and abstraction which is sculpture, in addition to the specific qualities of the painter's craft. He was as aware, however, of the differences between the arts as of their common properties. Had he not agreed with Cézanne on this point he would not have quoted with such approval his friend's remark: "It took me forty years to find out that painting is not sculpture." The latter art is one of line, plane, and volume; painting, when fully understood, gains its most characteristic effect through color, which is almost absent from the other craft. The pictures dating from the supreme last years of Renoir's life show the result of his work in the round with the actual materials and methods of the sculptor; yet instead of a loss of color we see his most complete understanding of it. And so there comes about such a final flowering of his paintings as even his greatest achievements had not prepared us to see. The design of his canvases had long since become almost infallible; his sense of form, developed over a half-century, now reached the point of effortless sureness as it merged indistinguishably with the man's unique sense of color. The understanding that color has a life of its own marks his ultimate separation from the realistic painting of his Impressionistic comrades and even from his early work. The next step in art, to be made by the generation that followed, would be the use of color and form in ever-increasing independence of the object represented.

Previous centuries had given to the world such giants as Titian and Rubens, whose overpowering achievements might at first lead one to think that painting had said its last word and that nothing could follow but decadence. This is far from the truth; in the centuries after these men the monumentality inherited from fresco or mural painting was replaced by the sensibility of the later work, by its rendering of intimate and poignant ideas unsuited to the broad treatment of the early masters, who had large expanses of plaster or rough canvas to cover. One may admire the frescoes of Piero della Francesca or those of

16. VENUS. Collection Jacques Seligmann & Co.

Raphael, and see this tradition of largeness continued in the work of Poussin or of Chardin; but for a Renoir, with his portrayal of the mysterious beauty of childhood, of summer skies, flowers, and the female nude, the world had to wait for the modern period.

Renoir would not have allowed his own constant development to be used as an argument for progress in art, an idea which was one of his aversions. There may be progress in scientific knowledge and techniques, but art reached heights in early Egypt which no later time has surpassed. In their externals the arts may change, and must change, since the world does not twice wear the same aspect; in their essentials the arts are changeless, and the great modern of yesterday or today continues in his own fashion the work of the ancients.

If Renoir added new luster to painting, he still "stays in the ranks," he stands with his contemporaries; but by his assimilation of tradition he joins the long ranks of the masters of the past. He also reaches forward to the painters of today, and his open-mindedness to new developments in painting, exemplified in his attitude toward Matisse, has been rewarded by the homage of the younger generation. This is especially evident in later paintings of Bonnard.

The statement that Renoir continued to hold a place of honor among the devotees of nonrealistic painting may need some enlargement here. Consider the advances he made during his career: in his early pictures the accidentals of a scene, its light and shade, its perspective and detail, are represented almost as they occur in nature; in his final period he arrived at an art in which even the most deeply felt impressions are only part of a harmony which might be called abstract. It was the inner rightness of the painting—the invention of the artist's creative mind—which became progressively of dominant concern. If the design of a canvas demanded some alteration of natural proportion; if the undulations of his linear rhythms demanded a displacement of some part of the body; if the flow and play of light and color demanded a relation of foreground and background elements which was not according to photographic vision; Renoir had little hesitation in giving the needs of his composition precedence over the materialism of fact.

Indeed, in his old age, he emphatically rejected the notion that visual accuracy was an important criterion of excellence: "As for me, when I stand before a masterpiece, I am content with enjoyment. It's the professors who have discovered 'defects' in the masters.... But those very 'defects'

17. STUDY OF A GIRL. Collection Rosenberg & Stiebel

18. PEASANT GIRL TENDING A COW. *Photo Durand-Ruel*

may be needed. In Raphael's *St. Michael* there is a thigh that is a kilometer long! And perhaps it wouldn't be so good otherwise.... Again, take Veronese's *Marriage at Cana*. If that picture were in true perspective, it would be empty; for the people in the background, who are as big as those in the foreground and who play their part so well, would be little figures indeed. In the same way, the floor does not go back according to the rules: and perhaps that's the reason why it's so beautiful."

And yet, Renoir, unlike so many artists of this century, never ceased to represent nature; from her he continued to draw his inspiration to the very end. "How difficult it is," he said, "to find the exact stage at which a painting should stop in its imitation of nature. What is necessary is that you get the very 'feel' of the subject. A picture is not a catalogue. I love pictures which make me want to stroll about in them, if they are landscapes, or to caress them, if they are women.

"Don't ask me if painting should be objective or subjective—I don't give a damn about such things. It makes me wild to have young painters come to me and ask about the aims of painting. And then there are those who explain to me why I put a red or a blue in such-and-such a place.... Granted that our craft is difficult, complicated; I understand the soul-searchings. But all the same, a little simplicity, a little candor, is necessary. As for me,

19. GIRL WITH A GUITAR. Collection Rosenberg & Stiebel

I just struggle with my figures until they are a harmonious unity with their landscape background, and I want people to feel that neither the setting nor the figures are dull and lifeless."

It was the spirit which he sought in his late compositions, the feeling of things, their essential character. Renoir was one of those men who, in Robert Louis Stevenson's line, live for "a thing not seen by the eyes." More than most men Renoir knew the joy of looking at the world; but he also knew that the province of art is one of sensations or ideas that can be rendered only by the specific terms and disciplines of a particular craft, whether it be poetry, music, sculpture, or painting.

We may outgrow ideas and even religions, but our pleasure in forms which are simple, expressive, and balanced remains constant. The art of primitive Europe with its pre-Hellenic sculpture, for example, or that of the early Mexicans, or of the American Indians, is in many aspects strikingly similar to the most modern arts of our day. When we say that a work is ancient or modern, European or American, we have arrived at no really important distinction. What we ask of any art is a personal and enduring vision, and what we find so richly presented to us in Renoir is an affirmation of all that is young and joyous and alive, a sense of order and balance in the world about us.

COLORPLATES

Painted 1864

MADEMOISELLE ROMAINE LACAUX

32 × 25¹/₂"

The Cleveland Museum of Art (Gift of Hanna Fund)

This charming portrait, painted at the age of only twenty-three, is perhaps the earliest signed and dated picture by Renoir which has come down to us. We know little of his early work; in dissatisfaction, it is said, he destroyed most of what he had painted between 1862 and 1866.

Renoir visited the artist's paradise of that time, the village of Barbizon, and here the little girl's family, on vacation, commissioned the artist to do her portrait. Already in this youthful work we see the traits which were to give Renoir his high place among the greatest painters of all time. The canvas transmits in an amazing way the alert energy of the sitter. One feels the artist's irrepressible love for people, his amazing facility for catching individual character, and his genius for endowing a canvas with the spirit of youthful femininity. Renoir's apprenticeship at porcelain painting is recalled in the lustrous color, and in the china-delicate pinks of the face and hands.

Like many a French artist before him, Renoir searches out the decorative grace of his subject: he is enchanted by the playful curves of the edging on the pinafore, and this becomes a theme which is varied in the contours of the hair, blouse, and skirt. Decorative enrichment led to the almost iridescent treatment of the background to the right; and again to the brilliant invention of the hands which rest on a cluster of flowers. Through passages of a luminous warm tint in the blouse, Renoir carries the eye upward to the head; the reds of the flowered background, the earrings, and the lips lead us down again to the main statement of the red theme in the flowers.

By devices of organization such as these, Renoir gives this picture its beautiful pictorial harmony; but its supreme beauty comes from its emanation of the very spirit of childhood.

Painted 1866

SPRING BOUQUET

41¼ × 31⅝"

Fogg Art Museum, Harvard University, Cambridge, Massachusetts
(Grenville L. Winthrop Bequest)

For all its apparent looseness, this painting has a precise structure. The flowers spill over into the lower left-hand corner in joyous profusion, with an asymmetry as free and as wild as nature. But at once the artist proceeds to counter this unbalance. To the right of the vase, Renoir has developed a heavy shadow area, rich in purples, shaggy in contour, and sharply contrasted with the pool of light below it. Especially important is the placing of the ruler-straight lines in the lower right side of the canvas. In the setup from which Renoir painted there was perhaps no such line as the diagonal which breaks into a zigzag in the extreme corner; yet something of the kind is patently necessary. If the reader will cover this diagonal, he will see that the composition becomes lopsided.

This leads to an awareness of the basic structure. The composition is held in two triangles: a lower one with its apex at the rim of the bowl under the flowers; and the other with its apex at the top of the bouquet.

The Impressionist technique had not yet evolved when Renoir painted this picture. The brushing here is bold, the pigment fat and rich, suggesting the influence of Courbet. And instead of a generalized effect of luminous color which will later *suggest* flowers rather than *depict* them, here the petals are separate and distinct.

Yet the canvas glows with light and color, indicating that Impressionism is just around the corner; there is something of that school in the feeling of the out-of-doors which Renoir has captured in this canvas. The texture of the flowers is marvelously rendered, and one is tempted to say that the very perfume of the flowers is there, too.

The silvery-blue tonality is enlivened by accents of black; the lovely sprinkling of yellow starlike blossoms is sheer delight. Renoir is at the threshold of his career, but his taste already is exceptional.

50

Painted 1867

DIANA

75¾ × 50½"

National Gallery of Art, Washington, D.C. (Chester Dale Collection)

Renoir tells that one day a man came to buy this canvas; they couldn't come to terms, for the prospective buyer wanted only the deer. The rest of the picture did not interest him!

The buxom lady became a Diana only through expediency. Renoir was a canny, realistic man as well as an artist. Here is what he said of this picture: "I intended to do nothing more than a study of a nude. But the picture was considered pretty improper, so I put a bow in the model's hand and a deer at her feet. I added the skin of an animal to make her nakedness seem less blatant—and the picture became a *Diana!*" He had his eye on the Salon of 1867; but the painting was rejected.

The canvas is one of the few on which Renoir employed the palette knife; it was at the time when the artist admired the thick impasto pigment and the prodigious realism of Courbet. The picture is clearly a studio piece: the lighting of the figure is harsh and the setting artificial.

Renoir is still trying to find himself, but his instinct and sensibility are sure. Again we have the cool silver-blue tones (one may recall that while Renoir paints cool at the start of his career and warm later, the reverse is the usual development). Throughout the painting there is a pronounced interest in varied textures, substances distinct in color, value, and rendering. Paint quality ranges from the smooth, unbroken sky to the enriched and dazzling flesh, then on to the more vigorously painted rocks, and finally to the broad, loose flecking of the foliage.

In a general way, the cool tones move from the upper left to the lower right corner, while the warmer tones are on the opposite axis. The bright band with which Renoir hoped "to make her nakedness less blatant" is part of the warm axis, but its lines follow the cool direction. There are many subtleties which the observer will discover: this is one of the great figure paintings of nineteenth-century realism, and its qualities are not easily exhausted.

Painted 1872

PONT NEUF

29¼ × 36½"

National Gallery of Art, Washington, D.C. (Ailsa Mellon Bruce Collection)

In this masterpiece of the painting of light, we have the happiest qualities of a bright summer day in Paris. How wonderfully Renoir has caught the vibrant many-sidedness of the city! In a most extraordinary way we are made to sense the large and simple compositional structure of the canvas, while at the same time we see the multitude of little details and contrasts of color and shape. The world is made of spots: buildings, windows, chimneys, flags, vehicles, statues, people; it is made of sky and clouds, light and air, of stone and water and foliage.

Despite the illusion of dazzling light, the picture is cool in tonality, even in the glare of the street. Yet within this larger simplicity, the whole palette is engaged: the yellows, reds, blues, violets, greens, blacks and whites. There is an incredible richness of hue in every section: a good example is the varied blues of the bridge structure in the lower right.

The picture is divided into three big regions, each with its distinctive color, silhouette, and type of variation. The sky—a light zone—is spotted with irregular clouds, finely graded as to size, shape, and luminosity. The architectural mid-zone is the darkest: on the left a complicated grid of beautifully varied horizontal and vertical lines; to the right the broader elements of bridge and water, with strong diagonals leading to the extreme lower right corner. The lowest zone, the street, is the lightest of all, and here the promenaders and their shadows, scattered and free, repeat the verticals and diagonals of the mid-zone. If the reader will turn the picture upside down, many of these relationships will appear more clearly.

Renoir daringly makes the ground much whiter than the sun-drenched sky; he deliberately accepts the glare in his eyes to paint what is perhaps one of the first realistically back-lighted landscapes in the history of art. He creates heat through coolness; the chaos of a busy street through order; and under his brush the grimy commonplaces of the city sparkle like jewels.

Painted about 1874

MADAME MONET LYING ON A SOFA

21 × 28¼"

Calouste Gulbenkian Museum, Lisbon

Renoir and Claude Monet were the two most daring innovators among the Impressionists and between them existed a close, life-long friendship. At the time this picture was painted, Renoir often stayed at the Monet house in Argenteuil, a small community near Paris, on the Seine. The two men painted many pictures there, often setting up easels side by side. As a result of their and other Impressionists' pictures, Argenteuil will have a lasting renown in the annals of art as a proving ground for Impressionist theories.

On one occasion Edouard Manet joined the two friends. He was working on a picture of Mme. Monet and her son, with Monet in the background. From time to time he glanced at Renoir, who had also set up an easel. At an opportune moment, he drew Monet aside and, making what is one of the worst wrong guesses in history, said: "You're a good friend of Renoir's. Why don't you tell him to give up painting? You can see he'll never get anywhere."

Amusingly enough, Renoir at this time was considerably under the influence of Manet's painting, which he admired for its terse clarity, its boldness of attack, and its elimination of shadow. The plate opposite is very Manet-like, and shows these characteristics.

It is really a study in instantaneous vision: in every respect the canvas exemplifies the attempt to strip painting to what is caught in a fleeting glimpse. Mme. Monet, at her ease on the couch, looks up momentarily from her reading; the artist, in a stenographic manner, notes the main color areas and shapes. There is almost no modeling, sharp contrasts of color and value establish features, hair, the details of the costume. The figure cuts across the canvas on a diagonal; one half is the thriftiest possible indication of pillows and wall, the other half is given to sketchy details and more assertive color. The picture has grace, air, vivacity, and complete assurance: a veritable tour de force.

56

Painted 1874

THE LOGE

31 × 25"

Courtauld Institute Galleries, London

"Beauty," said Stendhal, "is the promise of happiness." To Renoir, a simpler man, the words are synonymous: beauty *is* happiness. The promise and the realization are one. This picture is a hymn to the beauty of woman. It is an image of health, an exaltation of maturity, an idealization of togetherness. With what tact Renoir has placed the man in the background, covered half his face, and subdued the detail with which he is rendered! The woman is offered for full and rapturous gaze; her face and body and costume are more flower-like than the blossoms in her hair and corsage. She is a bouquet herself.

The man is Renoir's younger brother, Edmond, who worshiped him. The woman is Nini Lopez, one of Renoir's models at the time, "with a profile of antique purity." This goddess chose to leave Renoir shortly after and marry a tenth-rate actor.

"Black," said Renoir, "is the queen of colors." A typically Renoir choice, it plays throughout, lustrous, patterned, varied, uniting the man and the woman. The alternate stripes stream down from the woman's bosom and face, like a radiation from the glamorous flesh.

The picture is full of Renoir's best qualities. Every aspect, by plan or by instinct, is harmoniously developed for visual delight. Take, for example, the pairing of things: the man and the woman; the gold of her bracelet and the opera glasses; the double gold stripe below her hand; the two pink blossom clusters on her bodice; the two vertical stripes in the light drapery; the twin pinks of her face and the flower above it; the warm spots of his face and gloved hand; even, as a final flourish, the two splashes of black on the ermine at the edge of his shirt front.

Again and again these paired attractions are related through short lower-right-to-upper-left diagonals; contrasting diagonals form a major movement upward and into the picture through the positions of the railing, the woman, and the man.

One's eye keeps wandering back to that lovely face, its doll-like perfection set off by playful wisps of hair which keep this beauty from cloying. If painting can be compared with music, surely this canvas is Mozartian.

58

Painted about 1875

MOTHER AND CHILDREN

66⅛ × 41⅜"

Mother and Children (also called *The Promenade*) is a perfect expression of the springtime of Renoir's talent. While its forms and color promise the rich efful-gence of his later years, they remain discreet and almost shyly reticent. The scene is set in a park, where a mother guides her two daughters along a circling path. Together they form a triangular group set apart from the more distant figures. A hint of symmetry in the arrangement of the three main figures lends a note of decorative elegance. Yet the whole composition is developed as a series of off-balances. The mother and daughters are turned slightly to the right as they move along the walk. The interplay of the formal and informal thus introduced is continued in the relationship of the two small girls. They are dressed alike and similarly posed. But the younger sister at the left is more enclosed, with her hands hidden in a white fur muff. The elder is more self-confident, as, with a cocky turn of the head, she strides forward, holding her doll in firmly clasped hands. The lovely young mother urges them quietly along their way, her reas-suring hand touched to the shoulder of her smaller child.

To affirm the oblique movement of his portrait group, Renoir emphasizes the diagonal direction of the path. Partly to contrast with this lateral movement, and partly to balance the weighted left side of his composition, he has massed the other figures in the extreme upper right of the scene. The overlap of the older girl's hat and the dress of the woman seated in the middle distance serves as a transition to join the two groups. Elsewhere a screen of foliage serves to set off the figures by limiting the movement into the picture space.

The feeling of budding freshness conveyed by the figure is carried throughout the canvas by the tender color harmonies. The greens and pale lavenders of the landscape are echoes of the more intense turquoise and creamy white of the girls' costumes and the rich tones of their mother's dress and fur-trimmed jacket. The sharper contrasts of the doll's frock repeat in miniature the shape and color "theme" of the mother. But at every turn, the painter has avoided the obvious with mystifying ease. There is no archness in its sentiment, no banality in its sweet and fragrant spirit.

Painted 1875–76

TWO LITTLE CIRCUS GIRLS

51½ × 38½"

The Art Institute of Chicago (Potter Palmer Collection)

The popular theater (as distinguished from the dramatic)—the variety houses, the circus, the opera, the ballet—enchanted the progressive painters of Renoir's youth. It was an ideal subject for painters in revolt against the overblown and sticky themes, classical, historical, or literary, then favored by the ruling conservatives.

But there were more profound reasons for their attraction. The young Impressionists were interested in the spectacle of life: the look of things, their transiency, their effect on the senses; and they devised techniques of such brilliant felicity that subject and method were perfectly fused. The theater, the races, and picnics made ideal subjects: pure vision, lightness of spirit, freedom from involvement with meaning or cogitation.

This painting is a superb example. The two little girls are anonymous performers, pure and simple. But of them and of their momentary function, Renoir has created an image so lovely, so apt, that forever it will typify the spectacle show.

In a daring and fresh way, Renoir selects a high viewpoint which has the effect of placing the observer himself in the circus ring, to become part of the scene. The spectators in the background are cut, helping to preserve the completeness and portrait presence of the girls who pose for us.

Their figures stand out against a generally flattened ground; yet they are related to the distance by the lovely bluish tints of their costumes, repeated in the barrier behind them; the dark hair of one girl flows into the darks behind her; the reddish hue of the most distant ground receives its strongest statement on the railing, and then pervades the entire ground area in front.

This wonderful play of warms and cools, of lights and darks, is punctuated by the piquant grace notes of the orange balls and the golden yellows of the ribbons and trimmings on the costume and the shoes. The brushing is varied— a harmonious flow of little caressing strokes. The drawing of the figures provides long, graceful lines.

Much of the planning of this picture is due to the influence of Japanese art, which in the seventies was the great catalyst to the art of the Western world.

Painted about 1876

MADAME HENRIOT

27 × 21"

National Gallery of Art, Washington, D.C. (Gift of the Adele R. Levy Fund, Inc., 1961)

Who but Renoir could have given us a vision so rare and utterly charming? Who else could so miraculously capture this ineffable beauty and delicacy, the essence of femininity? The picture is so like a happy dream that it is a little hard to think of it as a portrait of a real and mortal being. Other painters—Renoir's idols, Watteau, Boucher, and Fragonard, for instance—have given us charm and delicacy; but almost no one has succeeded so completely in etherealizing a personality, while yet keeping it close and warm and human.

We know that the subject is something really seen: this is Renoir's magic. The figure is painted with a minimum of stylization, in colors atmospheric, diffuse, opalescent. With wonderful taste, Renoir gives fluidity to the lightness and bluishness of the whole, even though in order to do so he was obliged to sacrifice correctness, as in the drawing of the hands and arms, the neck and shoulders.

In its high-keyed tonality the whole canvas seems a radiance, its luminosity reaching a climax in the head. Through the very dark accents of the golden brown hair and the sparkling brilliance of the eyes, Renoir further assures the dominance of the face. Through the lovely, shadowless porcelain colors of the skin, the suppression of the drawing of the nose except for the accents of the nostrils, the striking emphasis of the eyes, Renoir makes of the features a serene, decorative ensemble. Many of these devices have determined a mode which has since been popularized by portraitists and fashion illustrators the world over. The superficialities can be copied; the poetic imagination which could conjure up such an image is Renoir's alone.

The sitter is Mme. Henriot, who often posed for Renoir at this time. She was an actress of the Comédie Française; through Renoir's pictures of her she has become the very image of womanhood at its loveliest.

64

Painted 1876

MOULIN DE LA GALETTE

51½ × 69"

The Louvre, Paris

"The world knew how to laugh in those days! Machinery had not absorbed all of life: you had leisure for enjoyment and no one was the worse for it." With this happy, wonderful picture before us, Renoir's reminiscence seems a pallid understatement. The canvas is so rich in attractions, so full of enchanting details, that it becomes nothing less than an affirmation of the goodness of living.

The painting celebrates the triumph of youth: the women are radiantly beautiful, the men as dashing and debonair as young blades ought to be. Renoir has become famous as a painter of the nude; but what painter has clothed the human form more entrancingly? And with unbelievable virtuosity, he has animated his figures with an amazing variety of postures and activities—bold, relaxed, eager, withdrawn, flirtatious—all of them graceful and natural.

There are bits of still life, shimmering patterns of the light fixtures, children—like the dainty blonde creature in the lower left—tucked in here and there. One even fancies that the buzz of voices, the shuffle of feet, and the gay dance tune are part of the composition.

This is one of Renoir's largest and most ambitious compositions; yet he was not to regard it as one of his best paintings. Despite its apparent crowding and turbulence, it reveals a studied organization. The triangular foreground group is related through silhouette and color to the group at the trees; and this group, through yellow and gold-brown tones, becomes part of a vertical unit which provides stability to the right of the canvas. The other side allows easy entrance into space over a ground dappled blue and pink—Renoir's way of creating the effect of sunlight and shadow without introducing neutral dark values. By emphasizing the verticality of the dancing figures through sharp color contrasts, Renoir echoes verticality again, and repeats it playfully in the posts in the background.

These are only a few of the linear relationships; varied curves set up another series of rhythms. Rich color is contrasted with plain, and each is developed into an independent sub-theme: reds, yellows, blues, greens, blacks. Light flickers across the scene, resting here and there for compositional emphasis. Subject and method have been completely integrated into a unity which is one of the great achievements in the art of painting.

Painted about 1876

GIRL READING

18 × 15"

The Louvre, Paris

One may imagine that Renoir, occupied on a canvas, suddenly caught a glimpse of the model relaxing in a sunlit corner of the studio, and, struck by the brilliance of her face in reflected light, snatched up another canvas and was well into the work before the girl realized what was going on. She had "a skin that takes the light," and for Renoir that meant painting, painting without respite.

What a dazzling thing he has made of this study! The light, reflected back from the book, makes the shadow of the face transparent, and Renoir there discovers a wonderful variety of fresh and subtle tints. The head glows as though lighted from within; and in the verve and spontaneity of the execution, in the casual ease of the pose, it recalls again the best in French eighteenth-century painting. The brush fairly dances a staccato beat, here laying down thick deposits and there mere washes. With rapid strokes of the brush, Renoir engages us in a red play; makes the gold flare up into the hair and repeats it on the book. With one dashing stroke the shape of the jaw is given. This parenthesis-curve is seen again in the lips, in the eyes and eyebrows, the book, and elsewhere. The major theme of the piece, however, is one not seen, and the word for it is insouciance.

The model was Margot, who is described as having curly auburn hair, sparse eyebrows, red lashless eyes, a wide nose, plump cheeks, her thick voluptuous lips curled in a scornful smile, and noisy and vulgar with her suburban accent. She is the dancing girl in the preceding plate, and was painted often by Renoir—always transfigured by his art. In 1881, this Montmartre demi-mondaine died of typhoid fever, and Renoir paid for her burial.

Painted 1876

NUDE IN THE SUNLIGHT

31¼ × 25"

The Louvre, Paris

Out of a riot of glittering brush strokes rises this stunning nude torso. "The most simple subjects are eternal," said Renoir. "The nude woman, whether she emerges from the waves of the sea, or from her own bed, is Venus, or Nini; and one's imagination cannot conceive anything better." The coloring and texture of the body indeed suggest rare sea-shell tints.

Evidently Renoir was delighted with the freshness and spontaneity of this sketch and chose to leave it so. The pearl-like shape and luster of the forms is the major theme, given in the round mass of the haunches and belly, the breasts, shoulders, head, and echoed in the roundness of arms and neck; and a dappled sunlight which plays across the arms and body sustains this motif. Nothing is allowed to distract from the fullness: notice how Renoir has virtually eliminated surface markings like the nipples and navel.

The girl is completely a thing of nature; only Renoir's recurrent bracelet and ring betray a note of feminine vanity. The grasses suddenly part, the scene becomes to our eyes a hubbub of color streaks, and before us is this unforgettable vision of a forest creature, a Rima to delight and trouble the senses.

Painted 1878

PORTRAIT OF A GIRL

25½ × 21¼"

Collection Mr. Jack Cotton, London

A striking thing about this picture is the adaptability and poetic assurance of Renoir's vision, which unerringly keys itself to the sentiment of the subject, the qualities of the persons or places he is painting.

Renoir protested the title, *La Pensée*, which became associated with this picture: "Why has such a title been given to my canvas? I wanted to picture a lovely, charming young woman without giving a title which would give rise to the belief that I wished to depict a state of mind of my model.... That girl never thought, she lived like a bird, and nothing more."

However this may have been, Renoir *was* caught by the pensiveness of the model on the chaise; wistful as she is, there is also a coquettishness, an element of sly appraisal, in her glance.

The girl inspired in Renoir the search for effects of reverie and mystery. The tones and contours are exceptionally soft and filmy; the loose, flowing hair and the diaphanous clothes help to make of her body a softened shell which is completely at one with her enveloping surroundings. Her colors — the warm hair, the lips, the flesh tones and their shadows—reappear in the meadow-like, freely spotted treatment of the chaise.

Renoir has searched out the nuances and mood of the subject, and has given us the charm and delicacy which come from deep and ready sympathy.

72

Painted 1878

THE SEINE AT ASNIÈRES ("THE SKIFF")

28 × 36¼"

The term Impressionism is really a misnomer for this type of painting. It was originally intended as a term of derision by academic minds, bewildered and antagonized by the sketchiness of the drawing, the shower of high-keyed brush strokes, the artless compositions, and the general feeling of evanescence. Yet Impressionism is anything but casual impression (as the picture *Madame Monet Lying on a Sofa* [page 57] may be described); it is, as may be seen in the present plate, a thoroughgoing, concentrated analysis of luminosity in terms of color. Everything else in the painter's bag of tricks is ruthlessly sacrificed to this preoccupation. And far from being "formless," as many critics still maintain, what the Impressionists devised was a new artistic form, amazingly consistent in every element, every intention.

Renoir was not often as thoroughly Impressionist as he is here. Cézanne is supposed to have said of Monet, the arch-Impressionist, "He is only an eye, but my God, what an eye!" Renoir was too much in love with people and with nature to be "only an eye" for very long. He worked as an Impressionist when the occasion suggested it, and he became one of the greatest; but life had many other songs for him, and this accounts for the variety of style and subject which we see in the plates preceding and following this one in our collection.

Painted 1878

MADAME CHARPENTIER AND HER CHILDREN

60½ × 74⅞″

The Metropolitan Museum of Art, New York (Wolfe Fund, 1907)

A brilliant and attractive woman, with great influence in the world of letters, art, and politics, Madame Charpentier became interested in the Impressionists in the late seventies, and particularly in Renoir. When he was commissioned to paint her—one of the most celebrated Parisian hostesses—at home with her children, Renoir seized the opportunity to make an impressive showing in the Salon of 1879, and thus encourage the reception of his work in circles that could afford to pay for it.

The plate opposite is, in a sense, the result of a wonderfully successful compromise. Here is that rare hybrid, a picture which meets the requirements of society portraiture, while at the same time engaging the artist fully in terms of his own personal creativeness. The patron, understandably, subordinates his interests in pictorial values to his hopes for an agreeable presentation of those qualities in his family which are precious to him. And certainly, as a charming revelation of a particular woman and her children, of her personality and the quality of her home, this picture is an unqualified success, and was so regarded at the Salon.

But it is also successful as a Renoir: full of grace and freshness, and arranged with a freedom rather unusual in a picture in which the various elements are not the choice of the artist at liberty in his own studio. Boldly, the artist concentrates attention on the children and the dog, who rolls his eyes in mock anguish over the weight on his back. In contrast to these attractions, through the placing of the figures and the perspective of the rug, the eye is led to the upper right corner, where there is a grouping of handsomely painted details, and an exit into the distance. To lend stability to the canvas and localize the vision, the prominent and richly varied black of the dress is placed in the center of the composition, taking the eye, by means of an animated silhouette, to the edge at the right.

Black and white, in the dress and in the dog, provide one major color motif; blue, another. To set them off, coral panels appear in the background, and the other areas of the canvas embody nuances of all the major colors.

Painted 1879

OARSMEN AT CHATOU

31⅞ × 39¼"

National Gallery of Art, Washington, D.C. (Gift of Samuel A. Lewisohn)

In this canvas Renoir has returned to one of his favorite spots along the Seine to explore again the beauties of the glinting water. True to Impressionist practice, the artist focuses on the glorious spectrum of outdoor light as it floods the landscape. With a technique that is at once dashing and rigorously controlled, he recomposes in paint his own soft vision of nature's radiance. Yet he remains sensitive also to the wealth of contrasting textures which greet his eye, and adjusts his brushwork to evoke not only the color but the "feel" of grass or water or windy, cloud-filled sky. That he accomplished this double task without sacrifice of internal consistency of technique, or loss of an over-all sense of artistic personality, shows how completely Renoir assimilated the Impressionist method to his own purposes. Almost casually he takes in the sweep of the landscape and the happy fellowship of the group of four sketched figures in the lower left of the composition.

The very presence of the boaters suggests another characteristic of Renoir's art. Though like the other members of the Impressionist group he painted landscape for itself, he loved to include his friends in his canvases. The rapport of the figures and their setting is complete. They are seen within nature, not distinct from it. They share the same qualities of texture, light, and color, as with easy informality they merge with the beauties of their surroundings. But their function is structural, as well as narrative. The accent of the group anchors the slashing diagonal of the vermilion boat, as it in turn counters the generally leftward movement of the river and its banks into the far distance at the upper left. The verticals of the figure moreover serve to support and clarify the horizontal elements of boats and farther river banks, which at once define and amplify the flow of the broad and sparkling waters. In one passage alone is the painter's intention perhaps unclear, where a blue-gray patch of sky at the upper left seems to move aggressively forward from its place on the distant horizon. Elsewhere the reconciliation of space and surface is extraordinary—particularly in reds. Their clear pattern of surface design never threatens the order of the space relationships, as the artist traces his color line through a range from scarlet to the palest pinks, and weaves it into the web of his contrasting tones. The result we all can see: a light-hearted world aglow with shimmering summer light.

Painted 1879

ON THE TERRACE

39³/₈ × 31¹/₂"

The Art Institute of Chicago (Mr. and Mrs. Lewis L. Coburn Memorial Collection)

Renoir's magic is intuitive. It is his very fear of the taint of formula which entices him into some of his most delicious and surprising inventions. This painting is a case in point.

In the most literal sense this composition derives from the traditional devices that often served as a point of departure for Renoir's basic arrangements. The young woman and child and the sewing basket together form a pyramid in the time-honored usage of generations of earlier painters. Also traditional is the balustrade, which sets off the figures from the background and lends a degree of architectural firmness to the otherwise complex irregularities of form. It was Renoir's gift, however, to be able to sustain his originality despite his respect for the accepted usage. That balance between tradition and invention is here masterfully exploited.

The figures seem immediate, caught in a particular moment. Their informality is brought out not only in the pose but in the extreme forward placement of the little girl; almost as a fragment, she appears to peek over the bottom limits of the frame. So too the basket of yarn and the green plant bucket to the extreme left. All seem crowded—almost prodigally—into an overflowing fore-ground. The terrace rail, set at a glancing angle, similarly disclaims the curse of dry geometry. Though it marks off a private corner in the shady park along the Seine, its thin and penetrable forms also provide a transition to the open tapestry of landscape behind the pert young woman and the winsome child. Trailing vines interlace with the rail and its slender metal supports, while at the right, the balustrade and potted plant become as one. In another order of relationship, the trunks of the trees in the middle ground echo with varying precision more regular man-made shapes. In ways like these, we can observe how the artist retained a sense of disarming simplicity in his canvas despite the complexity of his forms and color.

"I arrange my subject as I want it," the artist said, "then I go ahead and paint it like a child. I want a red to be sonorous, to sound like a bell; if it doesn't turn out that way, I put in more reds or other colors till I get it. I am no cleverer than that." The notion of arrangement is apparent; so too the deep trust in the worth of the thing observed. What is missing from this or any other "explanation" is the intuition that directs the selection and emphasis. Nowhere else has the painter found a red more "sonorous." But is it the red of the hats and yarn that is the directing force, or is it the opalescent tints of the flesh or the gleaming white of the girl's dress as it is contrasted with the deep blues of the other dress, or the filmy greens of the trees that stand open to the sky? No, it is not any one of these but all in their harmonious ensemble that have been the artist's tools in accomplishing his final aim: to conjure a blossoming glade, where a small girl and her friend are discovered quite by chance in a moment of quiet contentment.

Painted about 1880

HER FIRST EVENING OUT

$25^{1}/_{2} \times 19^{3}/_{4}''$

National Gallery, London

The rightness of Renoir's vision is a constant joy. How wonderfully he has caught the eager vitality, the wide-eyed freshness of this sweet and unspoiled child! She is dressed in her Sunday finery, with a bouquet of flowers clasped in her hands, spellbound by her introduction to the gay life of Paris.

How completely right is Renoir's concentration on her, painting the rest of the scene in flashing brush strokes that convey the dazzle of activity as it impresses the girl! Except for her, the canvas is a hubbub of movement and implied sound. In the boldest and simplest manner, he weaves the picture into a unity, through such devices as the daring repeat of the front contours of the girl's jacket in the curve of the partition beyond her, and in the variation of this line in her back. Daring also is the abrupt separation of near and far in the picture; but they are related by the partition which, though near, is painted in the colors and technique of the background. The color is held down in general value and intensity, but is altogether Renoiresque in its rich invention. Before such a picture, one is amazed at the way Renoir can catch and give reality to essences of the most elusive kind.

Painted 1880

NUDE

31½ × 25⅝"

Rodin Museum, Paris

For Renoir, painting was a free and natural act, uncluttered by theory and unadulterated by pretension. His study of a seated nude girl, once purchased by Rodin and now in the Rodin Museum, is an ingratiating illustration of this quality in his art. It has all the clearness and brightness of a sketch with no sacrifice of technical resolution in its forms and expression.

The subject is pleasantly familiar. She is young and clean and fresh. Though her gaze is casually averted, the whole character of her being is one of frank and open-eyed delight in the special powers of her femininity. She is a completely natural creature, untainted by the sins of Eve, her contemplation free from inner conflict. Her role as a model is not, however, one of an empty, passive drudge. For the artist and, one feels, for herself as well, her state reveals the strength of affirmation.

The pose sets much of the tone of the composition. For all its apparent informality, it composes into another of those grand pyramids which so often lend a stable breadth to Renoir's canvases. Within its simple containing shape the body assumes a relaxed and spontaneous posture as it turns easily on its axis. The dominating reserve of the pose carries into the execution of the picture. The individual forms are delicately understated. Although ironically this very work was once condemned for its drawing by a noted collector, who considered it unworthy of its author, it can now be appreciated as a remarkable example of his method. The prevailing softness of the forms finds clarity and substance in the selective focus achieved by precise, linear accents which describe with marvelous economy the turn of the neck, or the crease of flesh as the model's arm joins the shoulder near the breast, or the detachment of the right arm from the smooth drape that flows across her thighs. Aside from the area of the head, the interior modeling is but lightly suggested by a restricted range of pale tints alternating between warm pinks and faint rose shading off into the neutrals of the half-lights—they cannot be called shadows. The face is ruddier, with accents of red-orange in the small, pursed mouth and blooming cheeks. The flower-like glow of the flesh tints is sympathetically confirmed by the background, where a broadly indicated area of green, loosely suggesting foliage serves with the blue-black of the hair to bring out the brilliant transparency of the skin colors. Patterns of sharp yellow and muted pinks add a degree of light-filled depth to the generally flat handling of the setting. There seems little doubt that the artist had once intended to develop further these laconic suggestions of landscape, but it is clear too that he recognized with satisfaction the effectiveness of their present state of enduring promise.

Painted 1881

LUNCHEON OF THE BOATING PARTY

51 × 68"

Phillips Collection, Washington, D.C.

Not since the Venetian painters of the High Renaissance has the world seen such glowing opulence in painting. But whereas the Venetians generally found their inspiration in the myths and lore of ancient times, Renoir's genius transmutes the common occurrences of everyday life into Olympian grandeur. These young gods and goddesses are friends of the painter, persons well known in Parisian art circles at the time. Aline Charigot, a favorite model whom Renoir married shortly after this picture was painted, sits at the left toying with the dog; the other girl at the table is another favorite model, Angèle, a lady of colorful repute; Caillebotte, wealthy engineer, talented spare-time painter, who early began to acquire his great collection of Impressionist paintings which is now the pride of the Louvre—after a frenzy of opposition to the bequest in the nineties—Caillebotte sits astride the chair; the lady who so kittenishly closes her ears to a naughty jest is probably the actress Jeanne Samary, painted by Renoir many times; the identity of most of the others is known.

We have already mentioned Renoir's felicity in inventing graceful, vivacious poses—poses which always seem as though this is the way people ought to look. We have mentioned his knack for enlivening his canvases with piquant notes—a face which emerges unexpectedly, a play of fingers, bits of still life, bonnets, ribbons, beards, stripes, flowers. As is customary in Renoir's large compositions (and the Venetians'), one side of the canvas is rich in things big and near; the other side presents a view into the distance—in this case, a breathtaking piece of Impressionist virtuosity.

Foreground and background are related in part by the awning, which in its striping combines the hues of the foliage with the warmer tones of the group; its delightful serpentine edge echoes freely the curves in the group, and the breeze which flutters the valance sweeps also across the balcony. The feeling of animation is given in many subtle and striking ways: for example, the perspective of the balcony leads the eye to the upper right, but the open visual path into the distance offers an opposed attraction. And all the while the eye is cunningly led back, through relationships of color—the spotting of blacks, for instance; and through line relationships, over backs, across heads, or following edges of color or light areas. And within every detail, no matter how small or casual, what a wonderful enrichment! This one canvas alone would be enough to assure a painter immortality.

Painted 1881

FRUITS FROM THE MIDI

20 × 25⅝″

The Art Institute of Chicago (Mr. and Mrs. Martin A. Ryerson Collection)

This still life is at once like and unlike what we expect of Renoir; it is full of surprises. A certain ruggedness of brushing, bulk, and drawing suggests Cézanne's influence; it may even have been painted while Renoir was visiting his friend at L'Estaque. But the profusion—there are no fewer than twenty-four fruits of many different varieties—is Renoir's and not Cézanne's. So, too, are the textural interest, the abundance of color, the preoccupation with the look and feel of things; the sustained interest in the variety of highlights is Renoir's, and so is the discovery of subtle decorative qualities of the subject.

The scalloped ellipse of the dish is the most obvious statement of a decorative theme. As the eye goes around the outermost edges of the whole still-life group, it describes a larger variant of this theme. Inside the cluster of fruit, and in single objects as well—the peppers, for instance—further echoes are present.

Opposed to this development on the left side of the canvas is the color arrangement. The bright reds are mainly on the right side, and with the peppers as the corners, they form an almost perfect equilateral triangle. The dark purple-blues, the yellows and greens are reserved for the other side of the composition; but they are picked up by the fruits in the center of the triangle of red, as the red is echoed by the two globes at the left. Similarly, the purples are introduced into the pomegranates and in the cast shadows (which are quite unusual for Renoir at this period, again suggesting Cézanne). All the major colors are shot through the upper part of the background, and delicate nuances occur again in the tablecloth.

Other themes are in evidence: the globular, which is varied from the sphere to the egg shape; the elongated ellipses of the leaves; and the irregular, jerky movement of the stems, a variant on the white edge of the dish and a foil for the bursting fullness of the fruit.

90

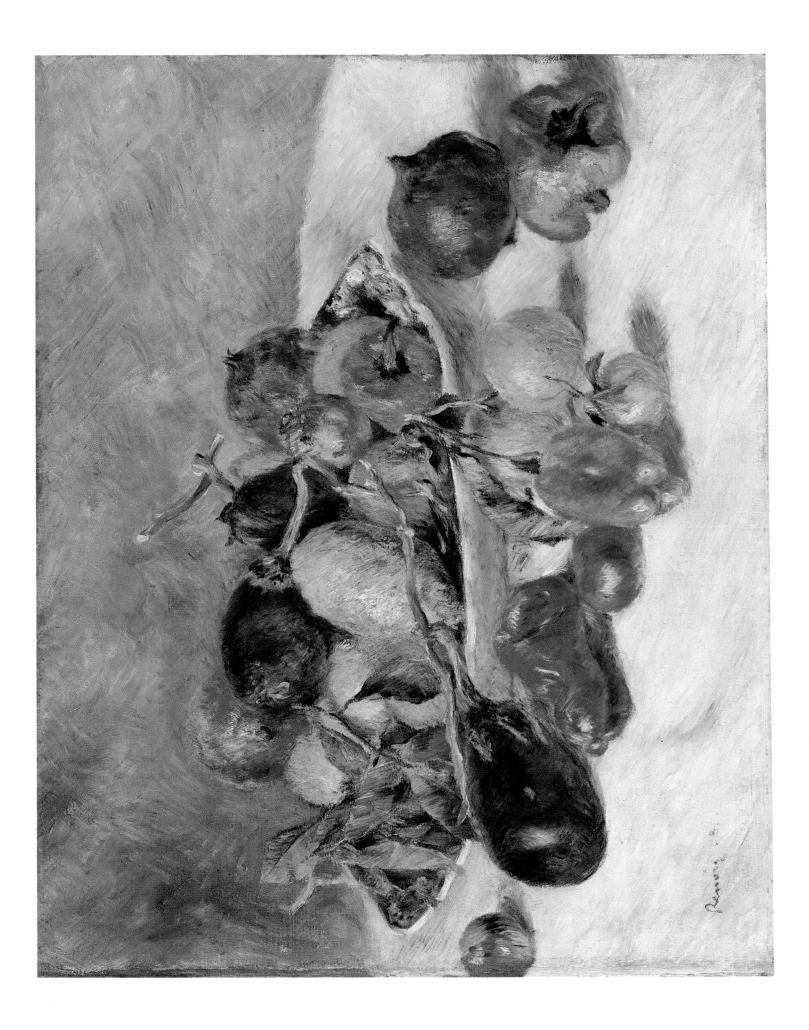

Painted 1882

CHARLES AND GEORGES DURAND-RUEL

25 ½ × 32"

Collection Durand-Ruel, Paris

In this straightforward portrait, Renoir gives us the likenesses of two sons of the famous dealer who early encouraged the Impressionists. There is a certain awkwardness about the composition, a "posed" quality, which is not present in Renoir's more successful works. The painting of the incidentals of the canvas is, however, superb. The lighter garments of the man at the left provide an excellent example of the way Renoir searches out the many variations within a single, general color; and the darker suit, more broadly painted, has a beautiful richness of hue. The foliage behind is freely brushed in, and yet conveys the feeling of profusion and depth. In our collection of plates, which summarizes the career of the artist, this is the last picture painted in Renoir's earlier Impressionist manner.

Painted 1883

DANCE AT BOUGIVAL

70½ × 37¾"

Museum of Fine Arts, Boston

No artist has ever been more aware than Renoir of the charms of his age, and nowhere else in his work does he celebrate his joy of life more beautifully than in the *Dance at Bougival*. His sense of identity with the scene is wholehearted and direct. These are the people he loved, whose simple joys he shared tirelessly and uncritically.

The girl and her partner are detached from their surroundings. Their relationship to each other is tender and unabashed. Even the most casual glance reveals that the man adores the girl. We see it in the tension of his hand as he restrains himself from squeezing her fingers too tightly, and in the carefulness with which he steers her into a turn, almost clumsy with affectionate concentration. And the girl, her body arched in the poised yet yielding pattern of the dance, turns her head and looks away—shyly delighted with the pleasure she inspires in her companion and herself. Typically for Renoir, it is the woman who takes on the greater identity, as her youthful, open face is framed enticingly by her red bonnet and brown bangs. By contrast, her admirer remains anonymous. The upper portion of his face is covered by the brim of a straw hat, and the massive bulk of his body serves but to emphasize the pliancy of hers, as even the red piping on her dress abstractly reiterates the pervasive rhythms of the dance.

The *Dance at Bougival* bears an interesting relationship with his earlier painting of the *Moulin de la Galette* (page 67). In fact, the dancing couple may well be an enlarged and refined version of the pair in the middle left of the previous composition, although the theme is one which attracted the artist to explore in more than one variation. Compared with the panoramic complexity of the composition of a few years before, Renoir has here turned to a concentrated focus on two figures, now enlarged to mural scale, as they loom above the spectator. The effect of their size is enhanced by the massing of the color into bold contrasts of red and yellow, pink and blue. The interior modeling, however, retains some sense of Impressionist handling and is thus consistent with the more fluid color and open textures of the rest of the painting, where a gay and sprightly crowd is engagingly but economically suggested. Three figures seated at a table to the left and fragments of a few others to the upper right suffice to conjure an atmosphere of outdoor festivity. But the chatter and music and the tinkle of glasses are distant sounds cast from the outer shell of the world where our lovers are joined to each other in the spirit of the dance.

Painted about 1884

GIRL WITH A STRAW HAT

21½ × 18½"

Private collection

In this face, smooth, plump, and solid as a tight-skinned apple, we have a new note. The meticulous drawing and firm modeling announce the artist's search at this time for a new discipline of draftsmanship in the tradition of the great Italians and in the art of Ingres, to which he turned with recently awakened eyes. Indeed, in the almost geometric restraint of the contours, with their firm, containing shapes, he rivals the greatest of his predecessors. The form is compact and solid. Yet it should be observed that it is not realized through the use of heavy shadow. Renoir is still the colorist, and it is because of subtle gradations of warm and cool tints that we get a feeling of roundness, of broadly bathing light and transparent shadows.

The reminiscence of Impressionism in the flesh tones finds fuller expression in the rest of the canvas. The basic color scheme exploits the opposition of complementaries—lustrous yellows and intense, saturated blues. The golden tones of the straw hat take on added glow as they contrast with the lapis-lazuli hues of the background. In an inversion of this color theme, the rich, warm browns of the hair enhance the coolness of the shadows on the dress as they in turn reflect the prevailing blue above. As brilliant as his orchestration of color is the painter's control over what could easily have become a confusing variety of brush textures: the subtle modulations of the skin, the silken smooth hair which cascades over the girl's shoulders to define the monumental pyramid of her figure, the fine lines—like engravings—in the treatment of the dress, the glittering straw of the hat, and the liquid sheen of the background, all contrive to focus our attention on the face of a lovely girl and make us aware of how a study like this can be at once precious and monumental.

Painted 1884–87

THE BATHERS

45½ × 66"

Philadelphia Museum of Art (Mr. and Mrs. Carroll S. Tyson Collection)

"After three years of experimentation, *The Bathers*, which I considered my masterwork, was finished. I sent it to an exhibition—and what a trouncing I got! This time, everybody, Huysmans in the forefront, agreed that I was really sunk; some even said I was irresponsible. And God knows how I labored over it!"

The abuse which greets any new direction in art seems bizarre in retrospect; and though Renoir had often been attacked, he could hardly have expected derision for this masterpiece of his *période aigre*. It represents an amazing summation of everything he had done and learned. Here are the color and luminosity of a great Impressionist; drawing which results from his admiration of Ingres and Raphael; the benefits of his researches into the clarity and simplicity of fresco painting; the playful grace of his adored eighteenth-century French predecessors; and, above all, that sweet ingenuousness which can exalt a bit of fun into something of Olympian grandeur.

The charms of the picture are not confined to the ladies alone: few painters in the history of art could succeed like Renoir in matching the natural allurements of subject with the allurements provided by his own taste and style. The refined, melodic drawing—a violin-clarity of line—is one of the great achievements of art. It is especially important in this picture to savor the decorative silhouettes and spaces which Renoir has so brilliantly invented: the arabesque made by the contours of rocks and feet is an example. The intricacy of shape- and line-play is daringly counterpointed against the uncomplicated, cameo-smooth appearance of the figures themselves. Renoir here uses flat, unshadowed lighting, which ordinarily subdues modeling; and yet, through delicate tints, he produces a luscious roundness in the bodies.

The spirited poses were derived in part from the seventeenth-century bas-reliefs of Girardon, at Versailles; the piquant faces, the vivacious gestures, the robust elegance are Renoir's. Cooks, house maids, *gamins*, shopgirls— Renoir paints them, and the world understands how it was that the gods of ancient times coveted mortal women.

Painted 1886

MOTHER AND CHILD

32 × 25½"

Ishizuka Collection, Japan

In this picture—surely one of the sweetest presentations of the mother-and-child theme—there is no hint of the mawkishness which could have come from a more literal rendering. Everywhere, Renoir has simplified. Style has been imposed upon his vision—a style which transforms every line, every color, every circumstance of space, shape, or light.

We observe, to begin with, the contrasted techniques of the human figures and the surrounding landscape. The latter is an accomplished piece of Impressionist painting, its atmospheric tones bleached yet warm, its forms and textures of a caressing softness. In contrast, the mother and her child are drawn with the precise contours of Renoir's "sour period," and their surfaces are painted in more flat, unbroken tones. Yet their chalky coloring—also derived from the study of Italian fresco painting a few years earlier—and many shapes and lines recall similar though weaker effects in the landscape; and one realizes how cunningly Renoir has brought into harmony the two antithetical manners.

As we study the figures, we observe what a brilliant draftsman was this artist who is more often celebrated for his color. The flattish areas of the white of the child's dress, the muted blue of the skirt, the tan of the jacket, and the flesh color—each has its distinctive contour and type of enrichment. Yet all are related: by edges which are continuous, by parallel directions, and by other correspondences—curves, points, and abrupt meanderings. Of particular interest are the intricate corkscrew silhouette of the child's legs and arm, and the related shapes of the mother's arm, breast, and head; and the way the breast and its nipple are almost exactly echoed in the drawing of the child's right thigh and hip, even to the point caused by the white skirt.

Renoir painted different versions of this picture of Mme. Renoir nursing their son Pierre, in 1885–86, and he took up the subject again after his wife's death in 1915, when he painted it in his loose, late style, and, with an assistant, sculptured it.

100

Painted 1888

AFTER THE BATH

$25^{1/2} \times 21^{1/4}''$

Private collection, Japan

Above everything else, Renoir is supreme in the history of art as a painter of the female nude. Some critics reproached him, not for prurience but rather for a certain "unseriousness" which led him to spend so much time painting nudes. Always he had the same answer, perhaps best summed up in some remarks he made to Vollard when they were talking about Boucher.

"I have been told many times that I ought not to like Boucher, because he is 'only a decorator.' As if being a decorator made any difference! Why, Boucher is one of the painters who best understood the female body. What fresh, youthful buttocks he painted, with the most enchanting little dimples! It's odd that people are never willing to give a man credit for what he can do. They say: 'I like Titian better than Boucher.' Good Lord, so do I! But that has nothing to do with the fact that Boucher painted lovely women superbly. A painter who has the feel for breasts and buttocks is saved!"

Renoir himself achieved not mere decoration, but, in the ancient sense of the word, "celebration"; and what he celebrated was the glory and wonder of the female form. To Renoir, as to William Blake, "the nakedness of woman is the work of God." He could never comprehend people who thought his painting frivolous, because for him it was almost an act of worship—a direct and reverent response to what he considered the supreme expression of nature.

It was his subject throughout the whole of his long life. In this example, from the end of the "sour period," Renoir foreshadows his later sculpturesque, monumental style; there is a new warmth that springs from fullness of color, a softening of edges, enrichment of textures, and a full development of forms. But more important than the changes of style is the uniformity of his vision: there is no hint of a personal relationship between model and painter. There is only wonder, unself-conscious and admiring.

Painted about 1890

GIRL WIPING HER FEET

25¾ × 21½"

Collection Mr. and Mrs. Grover Magnin, San Francisco

Renoir's "rainbow palette" is seen to perfection in this plate and the one following: his unique and rich orchestration of the whole gamut of colors. The mellow lushness, the range in any given area, from the most delicate yellowish nuances, through varied reds, greens, and blues, is something which no other artist has ever given us. One hardly thinks of pigment substance; the effect is rather of gorgeous, modulated light. And yet the quiet activity of the brush creates animation such as we sense when alone with the living things of nature.

In the midst of this Arcadian scene, on the banks of a curving brook, Renoir has placed a creature, ripe and full, aglow with animal health and the beauty of the flesh; without reservation his brush rejoices in the womanliness of his model. So the shepherdesses of classical literature must have looked.

Renoir's instinct for pictorial quality is seen in the way he takes an utterly natural pose, a most insignificant act, and dignifies it, gives it grandeur, and then climaxes it with the sweet unclouded expression of the face. The disarray of the garments, hinting at the delicious charms of the girl's body, is the poetic invention of a master.

The props are all familiar: we have seen this model in other pictures, a similar hat is worn by the girl on page 97, her garments are commonplace enough; and yet in all the variations the canvases are fresh and enchanting. A curious thing about this picture is the feeling of bigness which Renoir suggests; actually, the canvas is quite small.

Painted about 1890

IN THE MEADOW

32 × 25¾"

The Metropolitan Museum of Art, New York (Bequest of Samuel A. Lewisohn, 1951)

At this time, Renoir was emerging from his *période aigre*. He returned temporarily to Impressionism, but it was to an Impressionism wholly personal and original. With a new richness of color, and a new vivacity in his brush work, he paints thinly over light ground, creating a silken, undulant surface, like fine grass delicately agitated by a summer breeze. The effect is one of peculiar luminosity, as though, beneath the thin but complex colors, there was light and life. And although the figures are drawn with a fairly precise line—recalling his work of the eighties—we still remember pictures like this one for their color.

The painting has a simple basic structure—a kind of "X," the two diagonals of which contrast in every respect. From the lower right to upper left, a succession of bland colors moves from the hat in the corner, through the white figure, and into the distant landscape and sky; this diagonal lies both on the surface of the design and in depth. The other side of this "X" stays near the surface; it is made up of fuller colors, purples and coral. It follows the other girl's dress, from the lower left corner up into the trees, top right. Exquisite color contrasts are set up in all sections.

At the crossing of the "X," Renoir has placed a little bouquet of flowers in the hand of one of the girls. Again we are struck by the novelty—but complete naturalness—of the poses in Renoir's canvases; the girls have their backs to the observer, and they provide a human, but not personal, element.

Renoir's work in this period is sometimes criticized for "excessive softness of objects," as, for instance, in the foreground areas and in the trees. This would seem to be less a criticism than a misunderstanding. Renoir was after precisely that effect of softness and fluidity, that melting quality that permeates the atmosphere on a warm summer day; the pleasant lassitude of such a day is beautifully caught here.

Painted 1893

TWO GIRLS AT THE PIANO

27¼ × 23¼"

The Louvre, Paris

In this picture we have a detailed and felicitous account of the surroundings of French family life at the end of the nineteenth century, and yet it does not break down into a clutter of things asking for attention. We see the piano with its candle holders, the chairs, the tasseled draperies, and in the room beyond we get a glimpse of a stuffy but inviting confusion. We may study the costumes and the hairdo which the girls wear, and on the piano is a typical "old-fashioned" bouquet of the time.

This documentary character of Renoir's earlier work seldom conflicts with its qualities as art. The scene before us is bathed in a soft glow of light—its source and direction are not specific—and the rosy warmth helps us to join in the unself-conscious pleasure of the moment. This is a picture of leisure and relaxation and companionship. We feel this not only in the obvious indications of the subject, but more persuasively in the way the picture is put together.

The harsh, the rigid, and the angular are absent or suppressed; the gentle golden light which dilutes the colors brings them into the same family; and the ample forms which surround the girls seem to cushion and protect them. Yet Renoir here has given a surprising animation to a canvas which shows so little action. Our eyes are carried along sweeping lines and across forms which continue or stop one another. The foreground community of russet tones—the pillow on which the girl sits, the overstuffed chair with its music, and the piano —creates an arc moving to our right and toward the wall; the bodies of the girls lean forward in a similar but opposed arc; and this arc is restated in the drapery. The relation of the three arms and hands in the center of the canvas is a masterly variation on this theme. The girls, a lovely range of delicate pastel tints, are a beautifully organized group, typically Renoir in their vivacious postures and facial expressions. Typically Renoir also is the brilliant clustering of the heads and ornamental details at a point where many lines converge from above and below.

Painted about 1895

GABRIELLE WITH JEAN AND
A LITTLE GIRL

25½ × 31¾"

Norton Simon Collection, Los Angeles

Originally belonging to Cézanne, this canvas provides the most youthful portrait, in our collection, of Gabrielle. She is regarded with a special kind of affection by art lovers the world over: she seems like an old acquaintance, for her rosy-cheeked peasant face and her figure, luscious as ripe fruit, appear in many of the loveliest canvases of Renoir's later years. Shortly after the birth of the artist's second son, Jean (the famous motion-picture director), in 1893, Gabrielle was hired as his nurse—but not until she had met Renoir's usual basic condition for employment in his household: that she have a skin that "takes the light." How well she fulfilled this condition, the world knows.

Here, a girl of about fifteen, Gabrielle assumes, for the purposes of the picture, an aspect of womanhood only occasionally treated by Renoir: a mother-and-child relationship. It is an old theme in art; and if an immediate religious significance is lacking in this canvas, it is humanly more touching than many a religious picture.

Painted about 1895

THREE BATHERS

21½ × 25⅞"

The Cleveland Museum of Art (Gift from J. H. Wade Fund)

In the *Three Bathers* Renoir seems almost to pay a gracious compliment to the great decorators of the French eighteenth century, whom he continued to admire throughout his life. He made no secret of his affection for the lyric softness of Watteau or the adroit playfulness of Boucher. But this gesture (if such it may be called) is the more appealing for its independence, for Renoir's appreciation was too profound to allow him to fall into empty eclecticism. Though his terms were very different from Cézanne's, Renoir's goal in this respect at least resembled that of the Master of Aix: to bring tradition up to date; to relive it within the immediate demands of his own creative personality.

Although the subject appeared frequently in Renoir's *oeuvre,* this particular version of the bathers theme is one of his happiest achievements. Its gay spirit is infectious. Its limpid color harmonies and surging movement seem bafflingly effortless. Taken literally, however, the subject matter is trivial enough: a nude girl at the center of the composition teases one of her companions by holding out what appears to be a small crab. The second girl recoils defensively, while the others—a seated nude to the left foreground and two wading women in the middle distance—turn to watch the play. Fortunately this narrative incident was Renoir's starting point and not his end, so that what for a lesser artist would have remained a banal anecdote became instead an intoxicating vision of robust forms enveloped by a pearly morning light. It is almost as though a stage scrim has been dropped between the spectator and the performers, to enlarge and abstract the whole scene by the suppression of irrelevant detail.

Yet there is no lack of richness in the development of the forms, as with a caressing stroke of the brush the artist loads his paint onto his canvas. Nor is he unaware of subtle color differences as he adds a note of sharp red in a drape or counters the cobalts of the sea and sky with the blue-greens of the robe caught between the two cavorting figures. He differentiates the complexions of his three major figures to make each more individual than she may at first seem. And the colors of their glowing bodies are transposed into the landscape, where added notes of green suffice to convert them into a new context without loss of a sense of underlying affinity. Thus, with cajoling craft the painter disciplines his subject, training its rhythms and coaxing its colors into obedience. With patient yet hidden care he works out an exciting visual balance that lends breadth and vitality to a sense of carefree, simple abandon.

Painted 1898

ANEMONES

23 × 19½"

Private collection, Tokyo

Renoir loved to do flowers, and he used them often, spotted about in larger compositions and as subjects in their own right. Later on, he painted them as "experiments in flesh tones" for the nude. As may be seen in this collection of plates, he was sensitive not only to the color and obvious beauty of flowers, but also to their living character: they stand rigidly upright, they are twisted in growing, and occasionally they droop. With unfailing insight, he penetrates their *naturalness*.

There are many interesting technical things about this canvas, and we are, so to speak, brought up close to the painter's methods. The entire background of the picture may be seen to be washed in with pigment as thin as water-color. Inasmuch as the dark vase stabilizes the picture, the table top is almost entirely suppressed, lest its bulk overpower the composition: only a slight lightening of value indicates it. Yet, for the sake of controlled enrichment, the brush strokes here run in a contrasting horizontal direction.

The blossoms themselves are flicked in with quick stabs of the brush, which in places give them something of the effect of flames; in the upper, lavender ones, small whirling effects are set up. The play of stems and the ragged yellowish clusters help create that irregularity which Renoir loved.

Painted 1902

RECLINING NUDE

26½ × 60⅝″

The Museum of Modern Art, New York (Gift of Mr. and Mrs. Paul Rosenberg)

The great painters of the female nude have all given us their versions of this type of composition—the figure recumbent in graceful ease amidst settings bucolic, contrived, or intimate. One thinks of Titian's courtesans, Giorgione's goddesses, Boucher's eighteenth-century party girls, and also of Manet's brazen, tongue-in-cheek Olympia.

Like the panels of the earlier masters, this is not a study of the nude for the sake of representation primarily. It is more in the nature of lyric invention.

Here, Renoir has simplified both the drawing and the modeling of the figure to attain a surface which undulates in gentle swells, recalling bas-relief sculpture; and any strong projection from this flowing surface is suppressed. Thus, the angle formed by feet and lower legs is suggested only; so are the knee angles formed naturally in this posture by upper and lower legs; and, again, the angle where torso and thighs join. Anatomical details are not allowed to interrupt the soft movement of the surface, even at the expense of realism.

In harmony with this treatment of the body is the beautiful rhythm of long, graceful contours, sweeping but sensuous curves. The forms and contours of the foliage clusters and the distant hills echo the rhythmic motifs of the body.

The warm tints of the figure reach their fullest statement in the face and hair, and through the ruddy color behind the girl, presumably a garment, permeate the distant landscape. Conversely, the cools of the setting reach their maximum statement in the white cloth (so often the great painters contrast flesh with white) beneath her.

This type of indolently graceful figure became for a while a Renoir stock-in-trade: long-waisted, with the fullness of the hips placed quite low; full thighs; small breasts placed higher, and more widely separated, than is natural; full, chubby arms; a round head large in proportion to other details.

Renoir painted many variations of this sort of canvas and the best of them easily take their place beside their great ancestors in the past.

116

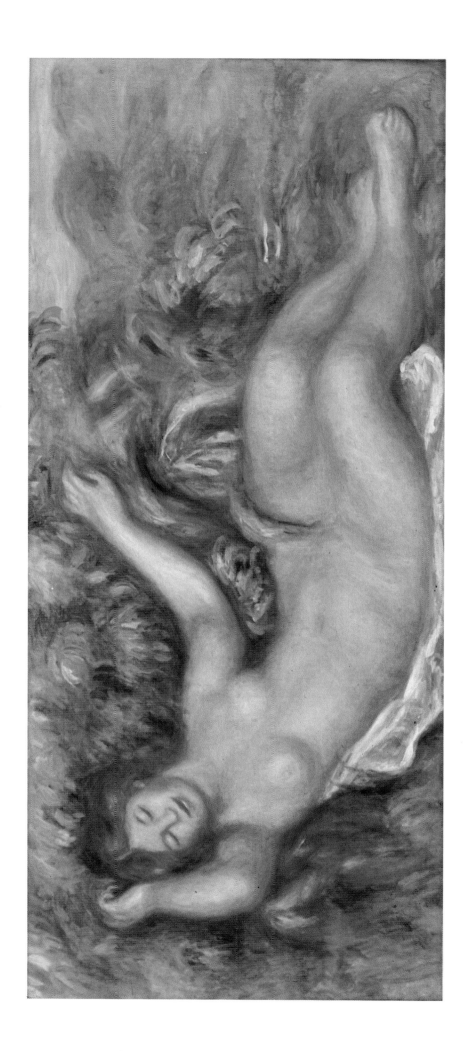

Painted about 1910

WOMAN AT THE FOUNTAIN

36 × 29"

Private collection, New York

The guiding principle of most of Renoir's later work might be summed up in the word "breadth." We have already seen how he broadened the poses of his portrait figures until they occupy the whole of the canvas; but this characteristic is much more than an obvious expansiveness. His color broadens out in the 1900's, and the small prismatic notes give way to simpler color areas in which a single hue is extended through all its nuances. The brush strokes themselves seem to have been made by a wider brush, and the stroke is less often a precise calligraphy: instead, there is something of a scrubbed effect.

The figures, as in the picture opposite, now are larger in their structure, and their curves are fuller and simpler: "correct" anatomy yields to amplitude. The figures suggest womanhood rather than femininity. The modeling, which in Renoir's earlier works often gave one the feeling that the other side of the figures might be flat, as in relief sculpture, now is generously rounded in form. Even the conception of representation is broadened. For Renoir has painted this figure not as seen from a single, fixed viewpoint: the head is side view, the torso almost front view, one breast front view and the other side view, and the legs quite flat and also side view. In this connection, one is reminded of archaic art with its disregard for fixity of viewpoint.

The landscape, too, becomes more expansive, composed of fewer individual elements, and in keeping with this broadening out, Renoir's late figures and backgrounds flow into one another through blurred or ambiguous outlines. It is almost as though he is painting nature as a single, all-pervasive substance; and this substance gains articulateness through its more or less vague resemblance to trees or ground in some places, while elsewhere it resolves itself into the freely shaped forms of the human figure; but in essence it is the same basic stuff.

It seems logical that Renoir, finding the earth and its inhabitants beautiful, should arrive at a conception of the oneness of all nature.

118

Painted about 1910

GABRIELLE IN A RED BLOUSE

21½ × 18"

*Fogg Art Museum, Harvard University, Cambridge, Massachusetts
(Maurice Wertheim Collection)*

This is the real Gabrielle, the handsome, big-shouldered woman who for many years served the Renoir household, as nursemaid to the children, as housekeeper, and finally as nurse and model to the aging and invalided painter. The simplicity of the portrait is strikingly effective: the absence of involved pose or composition, the restricted color scheme, the spontaneity of the painting itself—all echo the peasant directness of the subject. Renoir evokes here a sympathetic relationship between sitter and viewer, communicating a tender affection as he commemorates his devoted servant.

A comparison of this portrait with *Gabrielle in an Open Blouse* (page 33) shows us how Renoir draws out different aspects of the same subject, like variations on a theme. In the picture opposite, the inspiration is discovered in its source, the portrait concentrating on the personality of the sitter. This was the intention. In the picture on page 33, Gabrielle is not represented as a strongly individualized personality; what Renoir painted there is maidenhood, sweet and artless, and not yet quite conscious of its womanliness. Although painted only three years before the present picture, *Gabrielle in an Open Blouse* is a picture of youth and Eden innocence, as bright and cool as a cluster of wild flowers. The head thrusts forward a little stiffly, the bearing of the figure and the expression of the face suggest a timorous girl who is hoping to be told that she sat well—even though at the last moment she shied at complete nudity. The beautiful pink torso emerges from the pearly grays and satiny whites which give the blouse the effect of a translucent cocoon, and the simple color scheme is one of the most refreshing that Renoir ever invented.

In the portrait opposite, everything yields to the importance of the head, the warm colors and other details producing a feeling of relaxation and intimacy. *Gabrielle in an Open Blouse* charms through its elegance, poise, and the lovely substance of the pigment itself, through its adorable girlishness. *Gabrielle in a Red Blouse* is a winning picture of a sturdy, dependable worker and friend, whose patience and good humor are as rare as they are inspiring.

Painted 1911

SHEPHERD BOY

29½ × 36½"

Museum of Art, Rhode Island School of Design, Providence
(Museum Works of Art Fund)

This godlike shepherd boy, whose piping charms the birds out of the sky, is the son of a German industrialist named Thurneyssen. Renoir had painted the other members of the family, and when he came to this lad, he turned the portrait into a classical improvisation. As might be expected of a man who saw this earth as a paradise, and its inhabitants beautiful, Renoir always venerated the ancient Greeks. He had a keen understanding of their artistry, and one of the great regrets of his life was that he had never visited their homeland.

At the time when this picture was painted, Renoir had come to know Maillol well; he admired the tempered classicism of his sculpture, which was parallel to his own. There was a growing undercurrent of regard for the classical idiom in the early part of this century, and in Renoir we see the tendency at its boldest.

In this canvas, Renoir goes back to the Greeks via Impressionism, and via eighteenth-century France, when it was a conceit of the ladies and gentlemen of the court to play at being shepherds and shepherdesses.

It is worth noting also that Renoir almost never painted or sculptured the male nude. The torso is concealed in this picture; and the limbs, composed with a graceful harmony of repeated lines and angles, are those of his female figures. But in a poetic vision such as this it is of little importance: Renoir beautifully creates a rapport between the human figure and nature, and the scene is one of Arcadian tranquillity.

Painted 1914

MADAME TILLA DURIEUX

36¹/₂ × 29"

The Metropolitan Museum of Art, New York (Bequest of Stephen C. Clark, 1960)

In this one picture is seen the plentitude of Renoir's art: it has luxuriance without extravagance, and the composition, though basically simple, is of regal grace. The portrait presence of the sitter, however, is dominant: to look at the picture is like an encounter with the subject, a well-known German actress, wife of Paul Cassirer, the art dealer.

We think of the Baroque exuberance of Rubens, whose work Renoir studied, and of the Venetians. The figure has the noble amplitude we find in Titian, and in a manner which recalls his work, the pose—simplicity itself—is broadened and made to occupy the whole canvas.

And yet, Renoir has carefully avoided giving his sitter an unfeminine massiveness. He sections the figure in both vertical and horizontal directions, affirms the slenderness of the waist by converging lines, and suppresses the volume of the lower part of the body. The sectioning, with cunningly varied intervals and outlines, occurs throughout the canvas; in part it accounts for the richness-within-simplicity which is so unique here.

A long, flattened reverse curve, lower left to upper right, divides the picture into two almost equal areas, the right one rich with color, curves, and movement of the brush; the other, in which the head appears, is simpler in line and treatment. The same reverse curve is echoed in many parts of the canvas.

Most striking of all, perhaps, is that Renoir has produced the sensation of colorfulness with only two colors, a venetian red and a green, plus white. This enrichment of simple color is Renoir's particular genius. It includes here an elaboration worth noting: Renoir allows one color to be seen through another, and then spots the ensemble with sequin-like strokes.

Painted 1919

WOMAN WITH A MANDOLIN

22 × 22"

Private collection, New York

In this collection of plates we have traced Renoir's work as a painter. Here in a picture painted in the last year of his life, we see again that his pictorial resources were literally endless, and, at seventy-eight, he was still developing. The human presence of the model continued to be of central importance in his pictures, and here the womanliness is as pervasive as ever. And yet, as was pointed out in connection with the plate on page 37 Renoir's late work reflects something of the temper of the new painting of the day.

We are a long way from the canvases in which Renoir appeared as "the gentle, loving chronicler" of everyday incidents; we have left behind those paeans to glowing, sculpturesque bodies; those sensitive portrait characterizations. Now the human being is the *occasion* of the painting and not its ultimate subject. The canvases are primarily an ordering of ruddy colors, pigment textures, swelling lines, and patterns. All of these components have been broadened out; the colors and shapes begin to be separated from their contexts. Further development in this direction—difficult as it is to imagine that Renoir could become so "dehumanized"—would result in a coloristic abstract art. As it is, this canvas in some ways brings to mind Matisse's painting of the early twenties, although this picture is more directly sensuous.

Renoir's genius is constantly amazing: the number of his paintings runs into the thousands, and great as he was at any period of his long career, he went from one climactic stage to another, and always in the spirit of the time.